Lecture Notes in Intelligent Transportation and Infrastructure

The series "Lecture Notes in Intelligent Transportation and Infrastructure" (LNITI) publishes new developments and advances in the various areas of intelligent transportation and infrastructure. Merging theoretical foundations, practical applications, and forward-looking insights, LNITI provides a comprehensive understanding of both the state-of-the-art and the future prospects within this dynamic field.

LNITI is designed to be an inclusive platform that covers an extensive array of topics including, but not limited to intelligent transportation systems, smart mobility, intelligent logistics, critical infrastructure, smart architecture, smart cities, intelligent governance, construction design, data security, operational analysis, optimal route planning, digitalization, autonomous vehicles, the evolution of transport systems as well as green and sustainable urban structures. The series contains monographs, conference proceedings, edited volumes, lecture notes and textbooks. Of particular value to both the contributors and the readership are the short publication timeframe and the world-wide distribution, which enable wide and rapid dissemination of high-quality research output.

Proceedings published in the series are indexed by INSPEC.

All books, including proceedings, published in the series are evaluated by Web of Science.

Agnieszka Tubis

The New Paradigm of Risk in Internal Transport Supporting Logistics 4.0 System

Springer

Agnieszka Tubis
Department of Technical Systems
Maintenance and Operation
Wrocław University of Science
and Technology
Wrocław, Poland

ISSN 2523-3440 ISSN 2523-3459 (electronic)
Lecture Notes in Intelligent Transportation and Infrastructure
ISBN 978-3-031-64919-6 ISBN 978-3-031-64920-2 (eBook)
https://doi.org/10.1007/978-3-031-64920-2

This Springer imprint is published by the registered company Springer Nature Switzerland AG
The registered company address is: Gewerbestrasse 11, 6330 Cham, Switzerland

Preface

The development of Industry 4.0 (also known as the fourth industrial revolution) and the accompanying digitisation of processes in enterprises cause structural, organisational, and technological changes. This concept primarily concerned manufacturing processes in the initial stage of development, and researchers and managers concentrate their attention on creating intelligent factories. However, smart factories cannot be limited to smart manufacturing. Hence, there was a natural need to implement Industry 4.0 solutions also in the processes supporting the manufacturing process. This resulted in accelerating digital transformation, primarily in internal logistics processes, particularly in handling materials and finished products. These processes significantly affect the efficiency and continuity of production processes and are important for automated production.

Digital transformation in logistics has been going on for many years. It has become a significant trend aimed at improving information flows in integrated supply chains. In the initial transformation phase, digitisation was identified with the electronic version of documents exchanged between partners in the supply chain [1]. However, this concept is now perceived from a much broader perspective as the manifold sociotechnical phenomena and processes of adopting and using digital technologies in a more overall individual, organisational, and social context [2]. It introduces changes in the functioning of logistics systems comprehensively, as they concern [3]:

- Technical capabilities and technological infrastructure supporting material flows
- Organization strategies and directions for the development of logistics systems
- Human factor potential.

These changes enable monitoring of material flows in real-time, better handling of units, increased efficiency and effectiveness of logistics operations, accelerated and uninterrupted flow of information, and shortening the system's response time to emerging demands. To achieve these goals, the fourth industrial revolution must cover all company assets, including physical, information, and human resources.

Digital transformation and automation related to the development of the fourth industrial revolution affect the management of primary resources in the enterprise.

Therefore, it is also necessary to note that these resources are sources of hazards, the activation of which may cause adverse events that disrupt the operation of logistics systems. For this reason, it is necessary to analyse the impact of the implementation of Industry 4.0 solutions on the nature of all adverse events and, in particular, on their frequency, type, and consequences related to their occurrence. Assessing the existing risk in Industry 4.0 systems that support material flows (including internal transport) becomes a critical requirement for designing and operating solutions that are effective, safe, and resistant to interference. Its absence may limit the potential of implemented solutions or may also lead to situations that threaten the stability of internal logistics processes and the continuity of operation of entire supply chains. For this reason, developing a risk assessment method that considers the new dimension of logistics systems becomes a significant challenge for researchers and an expected tool by many managers who face the effects of adverse events that disrupt the operation of logistics processes.

The monograph aims to present a new risk assessment paradigm for Logistics 4.0 systems, particularly the internal transport system. This paradigm is concerned with the operational level of risk management. Technological development driven by Industry 4.0 solutions forces the need to look at systems supporting internal logistics by assessing their safety and technical reliability. It is an engineering approach that focuses primarily on the operation of mechanical systems. However, the fourth industrial revolution is also accompanied by organisational and process changes. These changes make managers focus primarily on achieving the goals set for logistics systems. They are assessed through the prism of performance measurement, i.e., indicators such as effectiveness, timeliness, completeness, and flexibility. The method presented in this monograph is intended to respond to these needs in an interdisciplinary way.

This book is composed of 6 Chapters. The aim of Chap. 1 is to introduce the reader to the most critical issues concerning the concept of risk assessment and management in modern organisations. Therefore, the concept of risk management, the most important concepts related to it, and the contemporary development trends intensely influencing the growth of interest in research on risk assessment methods in the scientific and industrial communities will be presented. The Author will also provide a rationale for choosing the research area described in the introduction of this book.

Chapter 2 presents the two dominant approaches to risk in contemporary scientific research and the proposed interdisciplinary approach. Therefore, the structure of the Chapter includes the investigation of business and engineering approaches to risk assessment. Based on the characteristics presented, the Author presents the need for an interdisciplinary approach to operational risk assessment in top technical systems.

Chapter 3 brings the reader closer to the changes in logistics systems. Therefore, it is necessary to describe two mutually complementary trends: the concept of Industry 4.0 and the digitisation of business processes. These phenomena and their impact on the contemporary development of logistics systems become the subject of the investigation presented in this Chapter. The obtained results allow to outline the background for the modern transformation we observe with the development of the concept of Logistics 4.0, which the Author also characterises.

The development of Logistics 4.0 and the automatic and autonomous solutions it uses in internal transport resulted in the emergence of new hazards and increased demand for analysis and management of existing risks. In Chap. 4, the Author presents the results of a systematic literature review in this area, considering two selected transport systems. The current research gap described by the Author in previous publications [4–7] will be confirmed.

Chapter 5 attempts to fill the identified research gap by presenting a new approach to risk assessment in Logistics 4.0 systems, focusing on material transport systems. The Author first compares the traditional material handling system (a human-technical system) with cyber-physical systems, described as cyber-human-technical (CHT) systems. Based on this comparison, the primary developmental changes and specific features characterising internal transport systems in Logistics 4.0 are identified. The Author developed a risk assessment framework for the system defined in this way, incorporating an interdisciplinary approach to risk identification and analysis, considering the specificity of the CHT system functioning in logistics processes. The proposed analytical approach was then implemented in a selected CHT system supporting internal transport, and the results obtained were presented in the form of a comparative analysis of the evaluation results carried out for human-technical and cyber-human-technical systems.

The last Chapter defines and discusses the main conclusions and directions for further research. The Author summarises the leading scientific and practical contributions of this monograph.

The research mainly used articles found in the Web of Science and Scopus databases. Articles in Polish were found using Google Scholar as a search engine. The Author primarily searched the relevant literature based on keywords, abstracts, and titles. Moreover, she also searched the articles for relevant references. The following principal terms and a combination of them were used for searching the literature: risk, risk assessment, risk analysis, risk identification, risk management, Logistics 4.0, Industry 4.0, Warehouse 4.0, digitalisation, industry revolution, autonomous vehicle, and automated vehicle. The literature overview was also supplied with relevant books and proceedings available for the Author (e.g., based on the AccessEngineering e-book resources database, Springer Link, or Taylor & Francis eBooks databases).

Although the Author has tried to give a reasonably complete survey, the reader may note that some papers still need to be included. The literature review was limited to publications available primarily in English and supplemented with publications in Polish. Thus, the Author apologises to readers and the researchers if they have omitted any relevant papers in the analysed research area.

This monograph is a response to the challenges currently faced by the management staff and engineers responsible for controlling enterprises' material flow handling systems. It is a valuable source of knowledge on the changes in modern internal transport systems and the associated risks. It was written based on research conducted by the Author, many discussions conducted during business training for logistics staff, and interviews and analyses prepared in cooperation with suppliers of Industry 4.0 solutions and clients implementing these solutions in their logistics processes. The results presented in the monograph combine the research and practical knowledge of the Author, who is an academic researcher, trainer, and business consultant. For this reason, the presented material will be helpful to many people, including logistics managers and engineers, graduates and senior undergraduate students, and scientists interested in the problems of risk assessment and management in Logistics 4.0 systems. In addition, this monograph will allow scientists to look at the concepts they teach more practically. Thanks to this, as part of the classes, they will be able to better prepare their students for the challenges that await them in modern logistics systems, which are currently characterized by high volatility and a growing risk of decision-making.

Wrocław, Poland Agnieszka Tubis

References

1. Matt DT, Pedrini G, Bonfanti A, Orzes G (2023) Industrial digitalization. A systematic literature review and research agenda. Eur Manage J. 41:47–78. https://doi.org/10.1016/j.emj.2022.01.001
2. Legner C, Eymann T, Hess T, Matt C, Böhmann T, Drews P, Mädche A, Urbach N, Ahlemann F (2017) Digitalization: opportunity and challenge for the business and information systems engineering community. Bus Inf Syst Eng 59:301–308. https://doi.org/10.1007/s12599-017-0484-2
3. Abdallah YO, Shehab E, Al-Ashaab A (2022) Developing a digital transformation process in the manufacturing sector: Egyptian case study. ISeB 20:613–630. https://doi.org/10.1007/s10257-022-00558-3
4. Tubis AA, Rohman J (2023) Intelligent warehouse in industry 4.0—systematic literature review. Sensors 23:4105. https://doi.org/10.3390/s23084105
5. Tubis AA, Poturaj H (2022) Risk related to AGV systems—open-access literature review. Energies (Basel) 15:8910. https://doi.org/10.3390/en15238910
6. Tubis AA, Żurek A (2022) The use of drones to handle internal logistic flows in the company from the chemical sector. In: Conference proceedings—CLC 2022, Bojnice, pp 107–112. https://doi.org/10.37904/clc.2022.4572
7. Tubis AA, Grzybowska K (2022) In search of industry 4.0 and logistics 4.0 in small-medium enterprises—a state of the art review. Energies (Basel) 15:8595. https://doi.org/10.3390/en15228595

Contents

Chapter 1
Introduction

Contents

1.1 The Risk Concept

The concept of risk has been widely known for centuries, but as an area of scientific research, the idea of risk assessment only emerged in the twentieth century. Depending on the scientific discipline, individual researchers will view risk differently. Thus, engineers, designers and constructors will view risk from a technological perspective; managers and economists will focus on its economic and financial side, while medical, environmental, occupational ergonomics and chemical engineers will adopt a perspective on health and life safety and environmental protection [1]. Therefore, numerous definitions of risk can be found in the literature. A significant body of research in this area belongs to Aven, who, in a series of articles, has published the results of his numerous analyses relating to the concept of risk, how it is defined, assessed and managed, but also the prospects for further development of concepts related to it (e.g. [2–7]). In one of his publications, Aven distinguished seven dominant approaches to understanding the concept of risk, which is usually referred to as [2]:

A. Tubis, *The New Paradigm of Risk in Internal Transport Supporting Logistics 4.0 System*, Lecture Notes in Intelligent Transportation and Infrastructure,
https://doi.org/10.1007/978-3-031-64920-2_1

- The possibility of an adverse event occurring.
- The possibility of undesirable (negative) consequences of the analysed event.
- The possibility of being exposed to negative consequences of an event whose occurrence is uncertain.
- The consequences of the action taken and the associated uncertainty.
- Intensity of consequences and uncertainty of action concerning something that humans value.
- The occurrence of specific consequences of the action under consideration and the associated uncertainty.
- Deviation from the reference value and associated uncertainty.

The analysis of the above approaches indicates that uncertainty is a critical concept in terms of risk. An overview of the discussions dedicated to understanding uncertainty in the context of the risk concept was compiled by Aven in [2]. COSO states uncertainty is "the inability to determine/calculate the exact probabilities or effect of future events" [8]. Van Asselt and Rotmans [9] distinguish two primary sources of uncertainty: (1) epistemic (due to lack of knowledge) and aleatoric (due to the variability of nature). Epistemic uncertainties may relate to a lack of measurements/observations and inaccurate or mutually exclusive evidence. In contrast, aleatoric uncertainties may arise from the variability of human behaviour, value diversity or the randomness of natural phenomena.

Uncertainty thus refers to the lack of knowledge (information) concerning the phenomenon under investigation (event, condition, scenario). This level can be assessed in terms of the difference between the information possessed and that required to make a decision. Referring to the level of knowledge held, researchers indicate that uncertainty can be [10]:

- measurable, in which case uncertainty can be quantified and is then referred to as risk;
- non-measurable, in which case it is referred to as strict uncertainty.

Real-world data collected or derived from a specific calculation model are used to quantify uncertainty. Therefore, probabilistic analysis is the predominant method to reduce random (representing variation) and epistemic uncertainty in prepared risk analyses [2]. However, it is worth noting that in analytical proceedings, uncertainty can take different forms and refer to various aspects concerning [11]: the event itself, the constraints on the parameters adopted in the model, the structure of the model, but also the effects of model inadequacy from indeterminacy or ignorance. A literature review indicates that the issue of uncertainty relating to the analysis model is becoming a critical research question [2]. Aven et al. [12] equate this uncertainty with the uncertainty relating to model error, which determines the quality of the analysis results obtained.

Based on these findings, risk is usually described in the literature as one of three approaches [2]:

- Risk as a measure of the probability of actions and their associated adverse effects.
- Risk as a combination of the probability and effect of a given adverse event.

- Risk as described by the triple (S_i, P_i, C_i), where S_i is the i-th scenario of a particular event/action, P_i is the probability of this scenario occurring, and C_i is the consequences associated with the occurrence of the ith scenario, for i = 1, 2, …, N.

It should be noted that risk in its fundamental dimension is always related to three elements describing it through the characteristics of the identified event, the estimated probability of its occurrence, and the associated consequences. Therefore, the risk is formally written in the form [2]:

$$R = \{E, P, C\} \tag{1.1}$$

where R—stands for risk; E—represents the events under analysis (adverse events, specific actions, scenarios); P—describes the probabilities of the analysed event E occurring; C—defines the consequences of the event E.

Besides identifying events, a critical element of the analysis becomes the estimation of the probability and the definition of a measure of the consequences of its occurrence. Recommendations on estimating the likelihood and valuing the consequences can be found in the guidelines formulated in ISO 31010:2018. According to this standard, there are three basic ways to calculate the probability of an event. These are [13]:

(1) Based on historical data, events or situations that have occurred in the past are identified. The data used should be tailored to the type of system, facility or process being evaluated, as well as to the operational standards of the organisation. Based on an analysis of the frequency of past occurrences of an event, the probability of future occurrences is estimated. However, it should be noted that in the case of a limited number of past occurrences of an event, the estimated probability may be incorrect.
(2) The use of predictive techniques (e.g. fault or event tree analysis) to forecast the probability of an event occurring. These methods are typically used when historical data are unavailable or insufficient. It is then necessary to derive probabilities by analysing the system, process or facility and the associated failure or success states.
(3) Without reliable quantitative data, risk estimation can be done by gathering expert opinions (expert opinions). This approach can be used in a systematic and structured probability estimation process. Expert opinions should be based on the relevant and available information (including historical data) specific to the system or organisation.

The final element that makes up the outcome of a risk analysis is the estimation of the consequences of an adverse event. Consequence analysis can range from a simple description of effects through quantitative models estimating the value of consequences to complex vulnerability analyses. The scope of the analyses depends on the decision-maker's information needs, which determine the range of impacts to be assessed and the stakeholders involved. Therefore, it is worth pointing out that consequence analysis can refer to [13]:

- the controls in place to address the consequences, together with all the factors influencing the consequences;
- the relationship of the consequences occurring to the original objectives;
- the direct consequences and the indirect impacts that may arise over a specified period according to the scope of the assessment;
- consider secondary consequences, i.e., those that affect related systems, activities, equipment, or organisations.

Risk assessment methods can use quantitative, qualitative and quantitative–qualitative models. The choice of model to be analysed depends on the availability of quantitative data, knowledge of the analytical techniques and tools used to assess the risk of the method, and the decision-maker's information needs. In the literature, the different approaches are classified into various groups. Qualitative techniques are used when experience and knowledge are available, while risks cannot be calculated (e.g. insufficient specific data). Quantitative methods are used primarily in more complex situations requiring much reliable information [8].

The dominant approach in research is to define risk in terms of an adverse event. The event is undesirable because it involves the activation of a specific hazard, the occurrence of which will negatively affect the achievement of stated objectives or cause a loss (e.g. financial, health, life). This approach is predominant primarily in engineering contexts, as indicated by the findings of Aven [4].

This is also supported by the fact that most risk management articles focus on measuring the effectiveness of different risk management systems, examining only the effectiveness of risk management in protecting against adverse events [14]. However, in economics and management, risk refers to events with a fixed uncertainty of occurrence (described by probability), the effects of which may be negative or positive on achieving a set objective. Therefore, ISO 31000:2018 defines risk as "the impact of uncertainty on objectives" [15]. At the same time, the approach proposed by the Committee of Sponsoring Organization of the Treadway Commission (COSO) [16] details that it is "the possibility that events will occur and affect the achievement of strategy and business objectives". As can be seen, these definitions do not specify the type of consequence resulting from the occurrence of an event (positive or negative) but only identify the significance of the event's impact on the achievement of the objective. However, it should be noted that ISO 31000:2018 details the scope of the proposed definition with some comments in the description, namely [15]:

- The uncertainty indicated may cause deviations from the expected state, either positive or negative.
- The objectives addressed by risk may relate to different aspects (financial, process, environmental, health and safety) and various levels of management (strategic, tactical, operational, project).

- Risks should refer to potential events and their consequences but also to combinations of them.
- Risk is often expressed as a combination of the consequences of an event (including changes in circumstances) and the accompanying probability of occurrence.
- Uncertainty is the state of complete or partial lack of information related to the understanding or knowledge of the event under analysis, its consequences and probability.

Different classifications and types of risk can be found in the literature, which is grouped according to specific division criteria. COSO recommends dividing the risk into four groups [8]: strategic, operations, reporting, and compliance. These categories are referred to by most of the classifications described in the literature. The most common risk categories described in the literature are shown in Table 1.1.

However, Razali and Tahir [18] note that most of the studies reported in the literature focus on financial, operational and strategic risk.

Table 1.1 Risk categories (based on [17])

Risk category	Characteristics
Strategic risk	• Influences the strategic direction and survival of the organisation; • Relates to macro-economic risks generated, for example, by government fiscal policy, and the impact of disruptive technologies such as the internet; • Involves poor business decisions and includes activities such as mergers and acquisitions
Business/financial risk	• Affects an organisation's business operations in terms of overall financial health; • Includes risks related to the market in which the organisation operates (known as market risk), as well as the ability to finance growth through borrowing (known as credit risk); • Is well realised, and many financial instruments and techniques support its management
Program/project risk	• Refers to initiatives related to implemented changes in the organisation; • In the case of the implementation of significant changes in the organisation, this risk can be closely linked to strategic risk; • Today, managing this type of risk is considered an essential managerial skill
Operational risk	• A broad category that includes failures in the area of any aspect related to an organisation's operations; • This includes operational management disruption, system or software failure, human error, process inefficiency, and procedural errors
Technological risk	• May relate to the launch of new technology products; • May refer to implementing new technologies into the organisation or its environment

1.2 The Concept of Risk Management

Although some authors claim that every business decision contains an element of risk [19], the literature distinguishes three decision-making situations occurring in connection with business activities [10]:

(1) Decisions made under conditions of certainty

When all the possible outcomes of the chosen actions are known, or the state of nature consists of only one element, the decision maker has certainty as to which state of nature will occur.

(2) Decisions made under conditions of risk

The decision maker is aware of the probability distribution of the various states of nature, which may result from theoretical assumptions or be an empirical distribution observed in the past. It may also result from the decision-maker's subjective assessment of the chances of particular states of nature occurring. They are the most common type of decision made.

(3) Decisions made under conditions of uncertainty

Occur when the decision-maker has no information about the probability of the realisation of states of nature or when, for example, a decision problem is being considered for the first time and previous experience cannot be used.

The majority of modern managerial decisions relate to decision situations 2 and 3. However, the objective adopted by managers should be to reduce the number of decisions made under conditions of uncertainty in favour of transforming them into decisions made under risk conditions. Indeed, for this group of decisions, it is possible to implement the concept of risk management.

Risk management is the formal process of assessing risk and making decisions based on its results. Risk assessment can relate to the various activities, processes, and resources used to achieve an organisation's objectives. Therefore, risk assessment results should appear at all levels and stages of planning and be an integral part of the decision-making process. It should be noted that the inclusion of risk analyses as a formal part of corporate decision-making processes dates back to the mid-twentieth century [20]. However, in the first decades, risk management was primarily associated with purchasing market insurance to protect individuals and businesses from various accident-related losses [21]. Therefore, risk analysis was seen as a tool dedicated to the financial and insurance markets. The continuous development of concepts concerning the transfer of risk through the purchase of insurance or derivatives to hedge financial risks is now an important research strand in risk management [22]. A second important research strand on enterprise risk is the approach derived from general management thinking on contingency planning. According to this concept, contingency planning should be part of preparing an organisation's strategy, which aims to identify activities exposed to disruptions and formulate solutions to limit losses caused by the activation of a hazard. Over the years, the concept of contingency planning based on risk analysis results has been continuously developed, and now, it is an essential part of the business continuity management approach [22]. A revolution in risk management came with the development of the concept of Enterprise Risk Management, which will be described in more detail in Sect. 2.1. This concept was presented in the document "Enterprise Risk Management—Integrated Framework", published by COSO in 2004 [8]. According to this approach, risk management aims to create a reference framework that allows an organisation to deal appropriately with the risks associated with its activities and processes [20]. This framework should describe an integrated approach to assessing, monitoring and controlling all risks, which should be identified at the strategic and operational levels associated with an organisation's activities. As noted by Pierce and Goldstein [23], the framework developed by COSO has been recognised by US companies as a good practice to guide risk management in the organisation.

Based on the ERM framework described by COSO, the ISO 31000:2018 standard was developed. This standard recommends an 8-step risk management procedure, as shown in Fig. 1.1. This procedure is the dominant approach to risk management in contemporary academic publications and industry recommendations. The risk assessment and management models currently described use the core framework of this procedure as the basis for proposed modifications.

The risk management procedure consists of five primary stages, which are interdependent, i.e. the effect of the previous stage becomes an element in the input of the next stage, and two supporting processes, which are implemented continuously.

The first phase of the process concerning identifying the context of the analysis is treated by some managers without due diligence, and it is a critical element of the analysis. This is because it is in this phase that the framework for the analytical and managerial investigation is defined. A crucial aspect of this phase is establishing the

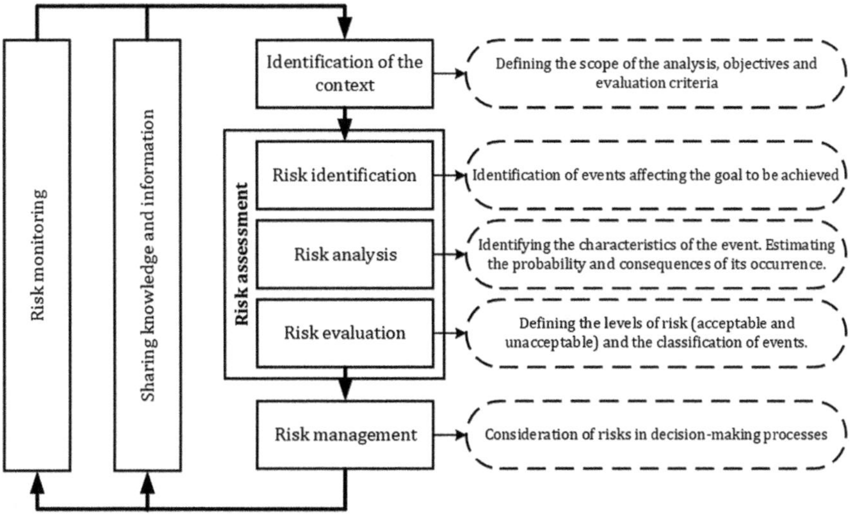

Fig. 1.1 Stages of risk management (based on [15])

scope of the analysis to be carried out, setting objectives and developing evaluation criteria. On this basis, techniques and tools for the analytical process and priorities for the risk management process are selected. The main stage is the risk assessment, which consists of three phases: (1) identification of events, (2) analysis of these events, and (3) risk evaluation.

The risk identification phase sets the scope for further analysis. Therefore, it is necessary to identify all the risks the organisation should manage, even if it has no control over their occurrence and impact [1]. Within this phase, analysts look for answers to the questions (Q1): What could happen?; (Q2): When and where could it happen?; (Q3): How and why will it happen? The next stage of the investigation is risk analysis. In this stage, the researchers estimate the risk parameters (primarily the probability and consequences of the assessed events). Qualitative, semi-quantitative, or quantitative methodologies are used in risk analysis.

The procedure aims to estimate the risk of occurrence of the analysed events in the form of a so-called risk indicator. These indicators provide input information for the next assessment stage—risk evaluation. Risk levels are first determined in this phase, guiding further risk management. The fundamental question the decision-maker answers is, therefore, what level of risk is acceptable or unacceptable to him or her. These two extreme ranges allow the appropriate priority to be given to actions aimed at risk mitigation. The risk indicators estimated at the analysis stage are then positioned according to the accepted limits of the designated acceptance levels. On this basis, priorities are indicated for actions to deal with the assessed risks. Risk management, as the last stage of the procedure, is already significantly linked to decision-making. This is because it concerns the selection of appropriate treatment alternatives to ensure that the risk is maintained at an acceptable level and that possible losses are avoided [24]. The monitoring and information-sharing processes support activities for the functioning of the risk management system; however, their implementation is critical for properly executing the primary phases discussed above.

1.3 Trends Increasing Interest in the Concept of Risk Management

Risk management in organisations has become increasingly important and famous, both academically and in business. Its development in recent years has been significantly influenced mainly by four phenomena, shown in Fig. 1.2.

1.3.1 Development of the Project Management Concept

Risk is now an integral part of the project management process, and its description is one of the crucial and mandatory elements of the project documentation

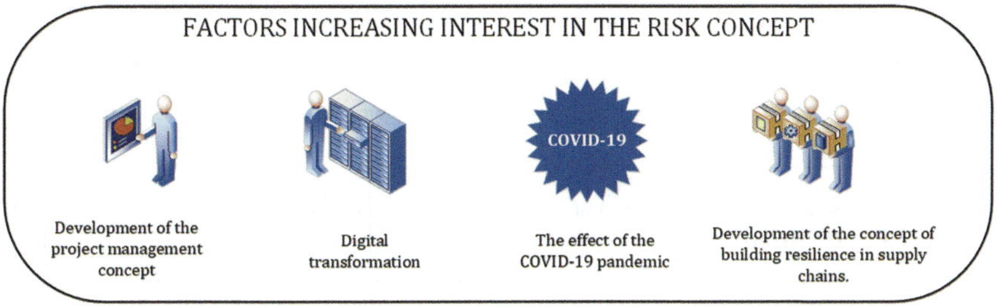

Fig. 1.2 Phenomena influencing the growing interest in the risk concept in companies

being prepared. According to the 7th edition of the PMBOK®Guide [25], the Project Management Institute defines risk as: "An uncertain event or condition that, if it occurs, has a positive or negative effect on a project's objective". Similar to ISO 31000:2018, the focus of the definition presented here is also on uncertainty that affects the achievement of the organisation's objectives. However, it is worth noting that the authors explain that there are many uncertainties in any activity, but only uncertainties that will affect the project will be considered risks. It is also important to note that uncertainty can refer to the event and the conditions under which the project is implemented. The developers of the PMBOK®Guide also point out that risk assessment should address both potential and actual opportunities or threats that will affect the project's objective during construction, commissioning or at the time of use.

As Elkington and Smallman [26] note, project delivery is less predictable in nature than routine business activities, increasing the uncertainty of decisions. Many authors emphasise that projects are characterised by greater organisational and technological complexity, which can generate additional risks and increased levels of risk [27–29]. Today, risk analysis has become a standard part of the project management process. As highlighted by many authors, risk management is one of the critical processes of project management [30]. This process should address every implementation stage, as it is essential to achieve the required project success rates [31, 32]. In addition, risk management helps project managers anticipate disruptions and delays that occur, which may result in the expected deliverables not being delivered within the required timeframe [33].

Project risk management should involve the same handling phases shown in Fig. 1.1. However, some researchers define it as a continuous and interactive process of identifying possible sources of risk at different stages of project development and execution [34]. For this reason, Elkington and Smallman [26] recognise that the most critical stage of risk assessment is the risk identification phase. This is because risks not identified early enough cannot be managed. Therefore, the risk identification should continue throughout the process [35]. Based on a literature review, Dandage et al. [36] identified categories of risks occurring in projects and their fundamental causes. These categories are shown in Fig. 1.3.

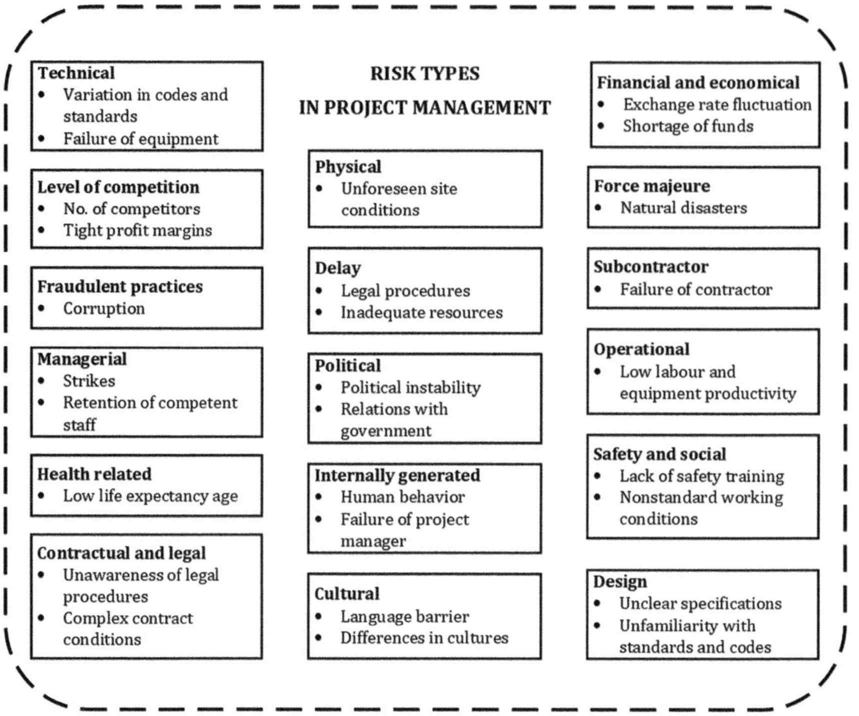

Fig. 1.3 Categories of risks and their causes (based on [36])

Because, in modern companies, most investments, process improvement activities, as well as the expansion of product offerings or the conquest of new markets, are based on project management, there is a growing interest among organisations in methods and tools for risk assessment and solutions aimed at dealing with risks effectively. This fact drives the continuous development of risk analysis methods and systems to support their implementation.

1.3.2 The Effect of the Covid-19 Pandemic

The COVID-19 pandemic was a humanitarian emergency caused by Coronavirus SARS-CoV-2 (later referred to as COVID-19), which began in Wuhan, China, in December 2019 and by the first quarter of 2020 had already affected the entire globe [37]. This pandemic is primarily a public health emergency. Still, it has highlighted the need for a transdisciplinary view of its impact, including from the perspective of global governance, technology development, and risk management and communication [38]. Indeed, the COVID-19 virus caused not only a health crisis but also an economic and geopolitical one [37]. The consequences of its transmission included significant disruptions to supply chains serving various sectors of the economy. In their research, Mankowski et al. [39] emphasise that the world has never faced a scale of impact such as that caused by the COVID-19 pandemic. The critical effects of the occurrence of this pandemic most commonly include [40]:

- Disruption of continuity of service due to changes in demand and supply of various goods.
- Reduced transport of goods.
- Changes in the conditions of production and distribution of finished goods and materials.
- Changes in consumer behaviour and preferences.
- Inadequate or limited emergency response plans of the organisation.
- High employee absenteeism; new systems are needed to schedule work crews.
- Reduced availability and delays in product deliveries.
- Difficulties in information management and inability to use historical data.

During COVID-19, the importance of analyses focused on risk assessment and management increased. Indeed, the detection of potential adverse events and the predictive determination of the consequences of their occurrence became critical. Risk analysis has become a valuable source of information on the incentives affecting businesses due to pandemics—their identification and understanding of cause-and-effect relationships provide the necessary knowledge required in managers' decision-making processes. The results of several studies indicated that proper information management was critical to the correct corporate response to the effects of the COVID-19 pandemic [40]. The availability of required information facilitated the early reaction from supply chains to emerging adverse events [41] while also contributing to the resolution of complex problems [39].

Critical to the survival of individual companies and entire supply chains became their response to emerging threats, occurring adverse events and their impact, the scale of which took most people by surprise. This response involved adapting ways of operating to the new socio-economic conditions, redefining objectives, and realistically estimating the time needed to achieve them. Maintaining organisational business continuity became the most crucial issue during the pandemic. The response of companies and supply chains to the impact caused by the COVID-19 pandemic has been analysed in the literature from different perspectives [40]. Some studies have focused primarily on companies' responses to emerging stimuli, such as the change in the pace of organisational processes or the frequency of product and service innovations generated during the pandemic [42]. Klockner et al. [43] identified five categories of business responses to the crisis caused by the COVID-19 pandemic, which are shown in Fig. 1.4.

Much research has focused on identifying effective and efficient strategies for responding to emerging socio-economic conditions. An interesting review study was conducted by Margherita and Heikkilä [44]. They aimed to identify response strategies that 50 large companies implemented to ensure their business continuity during a pandemic [44]. In the recommendations made by the authors, particular emphasis was placed on issues concerning risk assessment, at least on issues such

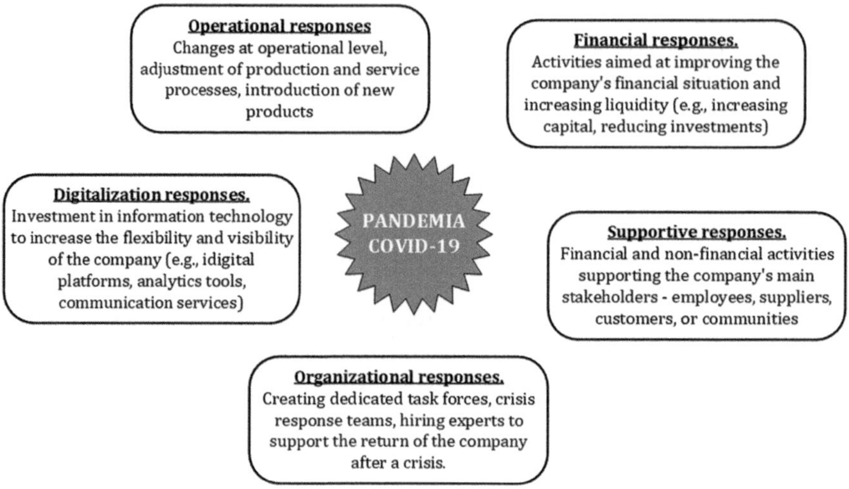

Fig. 1.4 Business responses to the occurring impacts of the COVID-19 pandemic (based on [43])

as the risk of infection inside and outside the organisation, the risk of limited technical and financial resources, and the risk of sales due to changing customer needs. Many authors have analysed specific supply chain management strategies and their impact on countering the adverse effects of a pandemic. An example of the results of such research is the publication [45], in which the authors evaluated the impact and effectiveness of strategies such as goal alignment, reliable leadership, use of integrated digital infrastructure, implementation of continuous learning and resilience development concepts.

Unfortunately, several studies suggest that plans to deal with the impact of pandemics on supply chains have been put in place with too much delay [40]. Remko [46] shows in his research that in the preparation phase to deal with the impact of a pandemic, businesses were insufficiently actioned, which influenced the delayed response of companies and their lack of resilience. This was mainly because the COVID-19 pandemic was a unique crisis with an unprecedented course and impact. Many researchers even considered this event to be a so-called "black swan" [47, 48], i.e. an event with a high level of uncertainty of occurrence (almost improbable) and impacts of critical importance for the economy and the world [49]. Therefore, the traditional approach to risk analysis proved insufficient for the new decision-making conditions managers face. Klockner et al. [43] also highlight in their research that the crisis management literature published before the pandemic focused exclusively on the different types of strategic responses to the crisis and their consequences. At the outbreak of the COVID-19 pandemic, there was thus a lack of interdisciplinary research on tactical responses, understood as short-term actions taken to implement solutions in response to an emerging crisis. It was only in the subsequent phases of the pandemic that researchers looked for solutions that would increase the efficiency and effectiveness of the actions taken to reduce the risk of adverse events.

Hohenstein [50], for example, pointed to the need to implement concepts such as agility, robustness, and learning from experience to improve the risk management process in times of crisis. The pandemic has also made managers realise that digital transformation is essential to their growth strategy and daily operations [51] rather than a solution dedicated to large corporations. Research presented in [52] demonstrates that the disruptions occurring in supply chains during the pandemic strongly motivated companies (regardless of their size) to implement digital business models. Giotopoulos et al. [52] also pointed out that implementing and accelerating digital transformation in organisations is widely considered a strategic priority related to building the resilience and competitiveness of modern businesses.

1.3.3 Digital Transformation

Digital transformation in various social and industrial areas has been observed for many years but was initially identified only by creating electronic versions of documents previously available in paper versions [53]. Nowadays, digitisation and its impact are much broader, as it refers to various socio-technical phenomena and processes related to implementing and using digital technologies in individual, organisational, and societal contexts [54]. Therefore, digital transformation is described as a change concerning the digitisation and digitalisation of processes in the economy, society, and different types of organisations. These changes take advantage of the continuous development of information technology, which enables the collection and processing of information from multiple sources to provide decision-makers with the required knowledge of processes and resources [55]. However, it is essential to emphasise that this change is not just about ITC systems supporting process execution but about implementing fundamental changes to the socio-technical-organisational system, including the organisation's strategy and culture [56]. Therefore, transformation must encompass the internal and external elements of the organisation and their interrelationships so that the effect of digitalisation is the expected improvement in the performance of the entire enterprise. Digital transformation must be multidisciplinary in nature to ensure the development and maintenance of the competitiveness of a given enterprise [57, 58].

Research on the areas that should be covered by digital transformation is often linked to models for assessing an organisation's digital maturity. These models make it possible to determine the current level of digital maturity in an organisation and steer its transformation in such a way as to achieve the desired goal. They can also benchmark and position the company in the market [59]. Numerous publications on digital maturity assessment models can be found in the literature [60]. In terms of identifying areas of digitalisation in an organisation, the results of the study presented by Hein-Pensel et al. [57] are noteworthy. The authors of this publication analysed 24 digital maturity assessment models and, based on the results, distinguished nine recurring assessment areas, shown in Fig. 1.5.

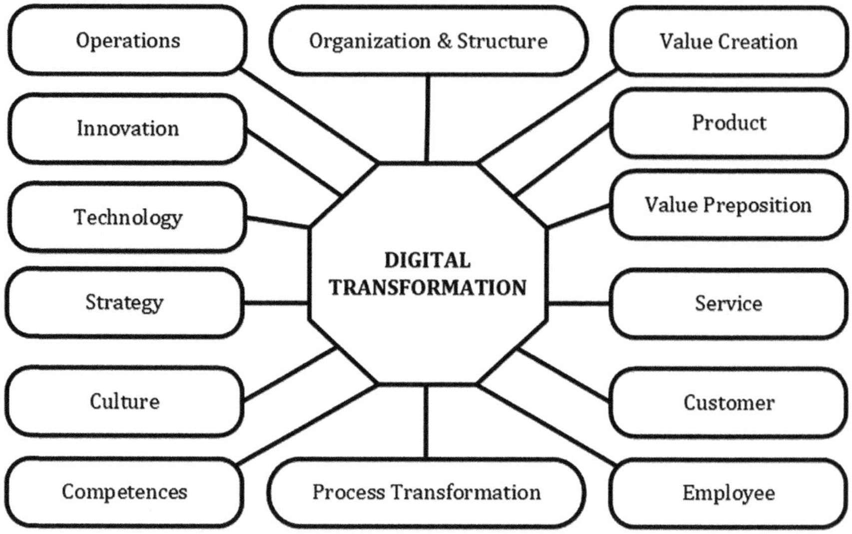

Fig. 1.5 Areas of digital transformation in companies (based on [57])

Many publications indicate that the acceleration of digital transformation has been significantly influenced by the development of the Industry 4.0 concept [61] (which will be described in detail in Chap. 3) and the COVID-19 pandemic [52, 62]. As the McKinsey Global Survey [63] of executives indicated, the pandemic accelerated digital transformation within companies by 3–4 years, both in external interactions (with customers and supply chain partners) and internal operations. At the same time, the share of digital or digitally-enabled products in companies' product portfolios offered to the market has increased impressively. This is mainly because many studies indicate that digitally mature companies are more resilient to crises [64, 65].

From the point of view of the topic of this book, it is important to note that digitalisation supports enterprise risk management processes, as modern organisations cannot manage risk without managing their data and knowledge. Indeed, since the goal of risk management is to understand the nature of risk and the effects of its occurrence to try to take proactive and preventive actions for mitigation, it is crucial to have access to comprehensive and timely information from various sources. Many authors (including [66, 67]) point out that the development of information technology has been a critical factor in supporting the development of the risk management concept by enabling organisations to:

- creating improved registries that collect data on different types of risk;
- modelling complex cause-and-effect analyses of different risks in an organisation;
- measure enterprise risk more accurately;
- a better understanding of the interdependencies between the various factors influencing the identified risks in an organisation.

As noted by Kang et al. [68], modern information and communication technologies provide opportunities to collect, store and analyse diverse risk-related data from heterogeneous data sources. For this reason, many organisations have started to use automated risk management frameworks to understand better their processes and the risks involved [69, 70]. An example of such a framework is proposed in [71], in which the developed framework provides comprehensive guidelines for the systematic capture, collection, analysis and monitoring of risk data, which may come from various data sources and be unstructured. Developing and implementing these frameworks supports managers in making intelligent risk management decisions. However, the organisation needs to reach an appropriate level of digital maturity to realise the full potential of these frameworks. Accordingly, the concept of "Risk Intelligence" has emerged in the research space, which can be defined as an organisation's ability to identify, predict, measure and assess risk based on collected data and experience [72].

However, to develop risk intelligence defined in this way, it is necessary to support managers with the right analytical tools to improve the decision-making process's efficiency, accuracy and timeliness. Therefore, there is a growing interest among researchers and managers in Business Intelligence (BI) and Data Mining (DM) tools [73]. BI systems are important in company decision support and risk management [74]. The system transforms raw data into useful information for risk assessment and management by accessing databases and using appropriate analytical tools and methods. At the same time, Liu et al. [74] emphasise that the implementation of BI in risk management must be based on an in-depth understanding of supply chain processes, management controls and sources of risk in individual companies and the supply chain. DM techniques are used at various stages of risk management to create proactive and reactive systems [70, 75]. They support gathering critical information on potential risk factors, their sources, impacts and inter-relationships [76]. Data Mining algorithms can also discover previously unknown but beneficial information and patterns of analysed phenomena from large data sets [77]. For this reason, these solutions are increasingly popular, and their use in enterprise risk assessment processes is growing [71]. Most commonly, DM solutions are used to extract information regarding [78]:

- prediction of risk events;
- discovery of risk patterns;
- relationship among risks and between risks and their triggering factors;
- classification of different items according to risks;
- clusters of different items based on risk factors;
- summarisation of the risk data;
- visualisation of the risk data.

However, it is important to note that the digital transformation has also resulted in the emergence of risks that were absent or very limited in traditional systems and are today becoming high-risk events. Such events include cyber-attacks, collisions between autonomous devices and people in a shared workspace, and dependence on digital technologies. Therefore, the digitisation and digitalisation of processes generate critical changes to the operating environment and the functioning of antropotechnical systems. This necessitates modifications in the various stages of assessing, handling and monitoring identified risks and in defining the context of the analyses to be carried out and their methods.

1.3.4 Development of the Concept of Building Resilience in Supply Chains

The growing interest in the concept of building resilience in supply chains stems from the high variability of the conditions under which modern chains operate, including the impact of the COVID-19 pandemic and other concurrent crises, the effects of which have been felt not only locally but also globally (e.g. the Russia–Ukraine war; the 2021 blockade of the Panama Canal; the winter of the century in Texas). However, researchers note that a formal definition is still lacking despite the observed development of the concept of building supply chain resilience [79, 80]. In the literature, this concept is usually identified as:

- property of being adaptive and capable of sustained [81];
- adaptive capability of the supply chain [82].
- ability of the supply chain to return/react/recover [83].
- inherent property to absorb shock and the ability to adapt to changes [84].

One of the more comprehensive definitions can be considered the proposal put forward by Ponomarov and Holcomb [83], who described resilience as the adaptive capacity of the supply chain to prepare for unexpected events properly, react quickly to emerging disruptions, but also to recover and return to the pre-disruption state while maintaining continuity of operations and control over structure and functions. Some authors even go a step further by indicating that the ensuing change should result in a return to the pre-disruption state and an increase and improvement in supply chain performance [85, 86]. Based on a literature review, Kamalahmadi and Parast [80] defined three supply chain resilience phases, shown in Fig. 1.6.

A critical element of the formulated definitions of supply chain resilience is the occurrence of a disruption. This concept is defined in most publications as an unplanned and unanticipated event that disrupts the normal flow of goods and materials in a supply network [79, 87]. From the point of view of the research being carried out, it is crucial to determine where this disruption occurs and the extent of its impact. When describing disruptions, some authors focus on the so-called vulnerable places in the supply chain, thus referring to the level of a single company or logistics system, while in the analyses prepared, the perspective of the entire network is adopted (e.g. [87, 88]). In some publications, the concept of disruption mainly refers to the occurrence of financial and operational risks [79], which can concern the level of a single organisation but also the entire network. For this reason, it is legitimate to clearly

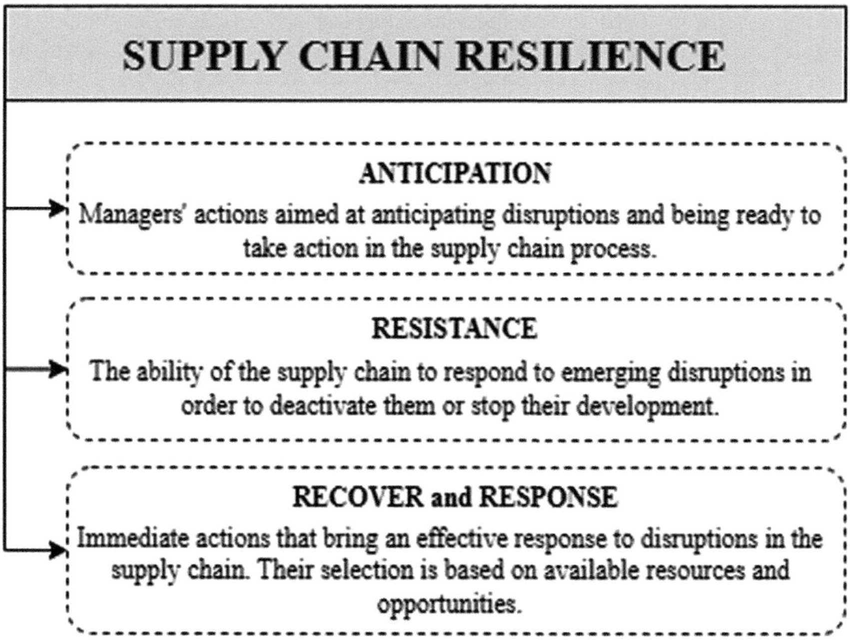

Fig. 1.6 Phases of supply chain resilience (based on [80])

define the type and nature of the disruptions analysed, as this will precisely define the scope of the research to be conducted and the level of risk analysis to be prepared.

Chopra and Sondhi, in their research on supply chain resilience, clearly indicated that managers should [89]:

- formulate a common cross-organisational understanding of the concept of supply chain risk and conduct stress testing;
- identify ways of adapting the overall approach to risk impact appropriate to the requirements of the individual company while considering the potential of the entire supply chain.

Consequently, when investigating supply chain resilience to disruptions, researchers often turn to the concept of supply chain risk management (SCRM) as a tool to support their analyses. Therefore, the development of the idea of supply chain resilience has led to a parallel increase in research interest in SCRM. This can be confirmed by Figs. 1.7 and 1.8, which show the number of publications identified by the document search in the Web of Science database. Only one restriction was adopted in the search procedure on publications that appeared from 2000 onwards, while the search itself was conducted according to the queries shown in Table 1.2.

Through risk assessment models, it is possible to examine and characterise the probability of a given disruption and the effects of its impact on the functioning of the entire supply chain and its single links. At the same time, many authors emphasise that

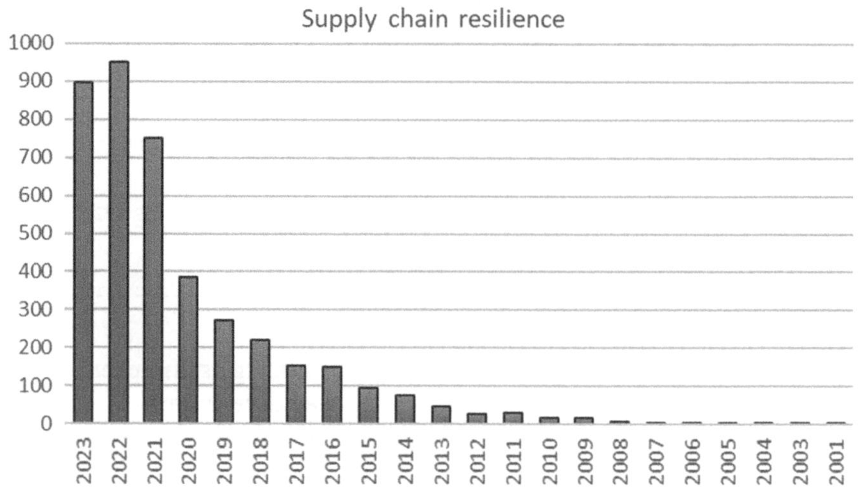

Fig. 1.7 The number of papers on supply chain resilience published in the Web of Science database

Fig. 1.8 The number of supply chain risk management papers published in the Web of Science database

Table 1.2 Search query for documents on supply chain resilience and supply chain risk management

TITLE_ABS_KEY (supply chain resilience) AND PUBYEAR > 1999
TITLE_ABS_KEY (supply chain risk management) AND PUBYEAR > 1999

appropriate risk management in the supply chain determines the sustainability of the relationship between links and the competitive position of the entire system [90, 91]. Kara [71] notes that increasing globalisation processes, shorter product life cycles and rising customer expectations increase internal and external risks. Emerging new risks to the functioning of supply chains arise from various sources, among which the literature mentions:

- demand uncertainty, disruptions to on-time delivery, political instability, currency fluctuations, dynamic changes in consumer markets, but also cyber-attacks, terrorism and natural disasters [92], and
- product characteristics, information systems, transportation, financial factors, manufacturing facilities, source process, make process, deliver process, return process, plan process, human resources, and external environment [93].

Therefore, there is a need to continuously examine the nature of disruptions occurring in the supply chain and the ways and extent of their impact.

The importance of the concept of risk management in building supply chain resilience is highlighted in numerous publications. Christopher and Peck [82] defined SCRM culture as one of the fundamental principles for ensuring supply chain resilience. The importance of this principle was further confirmed in the research described in [80]. In their further study, Christopher et al. [94] substantiated that creating a risk management culture and global sourcing reengineering are the two most popular approaches to mitigate the risks and ensure supply chain resilience. Soni et al. [95] also demonstrated that a risk management culture is a critical factor in determining the level of supply chain resilience to disruptions that occur. Waters [96] even argues that risk management should be an integrative part of the corporate culture in any organisation. Such conclusions are also confirmed in Mandal's research [97]. This author emphasises that risk management must be implemented by the supply chain leader and its partners. The results of this research confirm that a risk management culture should be implemented at all levels of supply chain management, i.e. it should not be limited to the strategic level only but should also apply to the operational level.

1.4 Purpose of the Monograph and Justification for the Choice of Research Area

The trends outlined in Sect. 1.3 have not only increased interest in the concept of risk management but have also changed managers' attitudes towards risk. Risk assessment and its use in decision-making processes is becoming a good practice that is applied not only by large financial corporations but also by SMEs. This is because they are the majority of actors in logistics networks supplying manufacturers in various sectors, and the fluidity of goods that flow along the supply chains they serve depends on their resilience to disruptions. At the same time, risk analysis is no longer the domain of strategic management. The disruptions recorded in supply chains in recent years have made managers realise the need to bring the concept of risk management down to the operational level. However, a change in the approach to establishing and operating a risk management system, which includes operating procedures, organisational culture, information systems and employees with specific competencies, is required. These changes must take into account the new conditions in which businesses are operating so that the risk management systems created to support the processes of identifying, assessing and dealing with risks that are appropriate to the economic, social and technological situation.

The risks associated with the operation of supply chains, as described in the literature, focus primarily on the collaborative aspects of the partners, and the building of resilience is based on the development of partner relationships, improved communication between links and increased flexibility of service, the effect of which is to maintain the continuity of the entire supply chain. However, the resilience of chains is strongly dependent on the reliability of the technical logistics systems handling the flow of goods and their operational readiness to perform their tasks. Industry 4.0 and the associated digitisation and automation of logistics operations have significantly influenced the changing nature of the risks associated with executing processes within internal logistics. Implementing this concept in logistics processes has a wide-ranging impact not only on the technical systems used but also on the human beings who play a new role in them [98]. Many publications emphasise that the implemented solutions of Industry 4.0 affect changes regarding socio-environmental norms, the labour market, the required qualifications and skills of the workforce and, therefore, the setting of new educational priorities [99, 100]. This necessitates the formulation of new guidelines and frameworks for creating risk management systems and developing dedicated rules for their operation that align with the requirements of Logistics 4.0.

The comprehensive literature review conducted, based on publications registered in the two largest journal databases (Web of Science and Scopus), indicates the existence of a significant research gap in methods for assessing and managing risk at the operational level in modern companies implementing Industry 4.0 solutions. This gap concerns, in particular, risk assessment methods for complex cyber-human-technology systems operating logistics processes in enterprises, the results of which would provide a basis for building the resilience of Logistics 4.0 systems. In the following chapters of this book, particularly in Chap. 4, the Author will prove the existence of such a formulated research gap.

At the same time, the industrial research carried out by the Author proves that many companies, when deciding to implement Industry 4.0 solutions, are not well prepared for this process. Managers do not have full knowledge of the risks associated with operating complex cyber-physical systems and their collaboration with employees. Thus, they cannot consider them in their decision-making processes during the planning phase. Focusing too much only on the technical aspects does not allow them to see other risks affecting the efficiency and effectiveness of the implemented processes. Therefore, managers are increasingly looking for solutions that will allow them to comprehensively (holistically) and multi-variantly assess the risk of possible adverse events occurring in various phases of the realisation of processes for which they are responsible. This is because most of them are aware that the implementation of Industry 4.0 changes not only the technical side but also the organisational and cultural side of the processes subordinate to them, and the results of the risk assessment can help them better prepare not only the system but also the people to work in the new operational conditions.

Therefore, the monograph aims to present a new risk assessment paradigm for Logistics 4.0 systems, particularly the internal transport system. This paradigm is concerned with the operational level of risk management. It combines two approaches to risk assessment for anthropotechnical systems supporting process execution in modern companies—the business approach and the engineering approach. In the book, the Author proposes a new approach to risk assessment in Logistics 4.0 systems based on the author's concept of an interdisciplinary process approach to risk assessment, digital transformation and automation of logistics processes, including internal transport in material handling. The proposed method is concerned with risk assessment at the operational level. However, the framework for proceeding must be embedded in the company's strategy to provide an effective and efficient mechanism for implementing the formulated process assumptions. It is also worth noting that the analytical approach adopted in the proposed method assumes that the primary objective of implementing an operational risk management system is to support the decision-making processes of managers in the enterprise. This approach aligns with the decision analysis manifesto presented by Howard [101].

References

1. Lu J, Jain LC, Zhang G (2012) Risk management in decision making. In: Lu J, Jain LC, Zhang G (eds) Handbook on decision making. Intelligent systems reference library. Springer, Berlin, pp 3–7. https://doi.org/10.1007/978-3-642-25755-1_1
2. Aven T (2016) Risk assessment and risk management: review of recent advances on their foundation. Eur J Oper Res 253:1–13. https://doi.org/10.1016/J.EJOR.2015.12.023
3. Aven T (2012) The risk concept—historical and recent development trends. Reliab Eng Syst Saf 99:33–44. https://doi.org/10.1016/j.ress.2011.11.006
4. Aven T (2010) On how to define, understand and describe risk. Reliab Eng Syst Saf 95:623–631. https://doi.org/10.1016/J.RESS.2010.01.011
5. Aven T, Reniers G (2013) How to define and interpret a probability in a risk and safety setting. Saf Sci 51:223–231. https://doi.org/10.1016/j.ssci.2012.06.005
6. Aven T, Renn O (2009) On risk defined as an event where the outcome is uncertain. J Risk Res 12:1–11. https://doi.org/10.1080/13669870802488883
7. Aven T, Krohn BS (2014) A new perspective on how to understand, assess and manage risk and the unforeseen. Reliab Eng Syst Saf 121:1–10. https://doi.org/10.1016/j.ress.2013.07.005
8. Committee of sponsoring organizations of the Treadway commission: enterprise risk management—integrated framework
9. van Asselt MBA, Rotmans J (2002) Uncertainty in integrated assessment modelling. Clim Change 54:75–105. https://doi.org/10.1023/A:1015783803445
10. Tubis A (2018) Operational risk management method in road transport. Oficyna Wydawnicza Politechniki Wrocławskiej, Wrocław

11. Spiegelhalter DJ, Riesch H (2011) Don't know, can't know: embracing deeper uncertainties when analysing risks. Philos Trans R Soc A: Math Phys Eng Sci 369:4730–4750. https://doi.org/10.1098/rsta.2011.0163
12. Aven T, Baraldi P, Flage R, Zio E (2013) Uncertainty in risk assessment: the representation and treatment of uncertainties by probabilistic and non-probabilistic methods. John Wiley & Sons
13. ISO 31010:2019 Risk management—risk assessment techniques
14. Andersen TJ (2008) The performance relationship of effective risk management: exploring the firm-specific investment rationale. Long Range Plann 41:155–176. https://doi.org/10.1016/j.lrp.2008.01.002
15. ISO 31000:2018 Risk management—guidelines
16. Committee of sponsoring organization of the Treadway commission: enterprise risk management—integrating strategy with performance
17. Kremljak Z, Kafol C (2014) Types of risk in a system engineering environment and software tools for risk analysis. Procedia Eng 69:177–183. https://doi.org/10.1016/j.proeng.2014.02.218
18. Razali AR, Tahir IM (2011) Review of the literature on enterprise risk management. Bus Manage Dyn 1:8–16
19. Lam J (2014) Enterprise risk management: from incentives to controls
20. Dionne G (2013) Risk management: history, definition, and critique. Risk Manage Insurance Rev 16:147–166. https://doi.org/10.1111/rmir.12016
21. Harrington S, Niehaus GR (2003) Risk management and insurance. McGraw-Hill, New York
22. Dickinson G (2001) Enterprise risk management: its origins and conceptual foundation. Geneva Pap Risk Insur Issues Pract 26:360–366. https://doi.org/10.1111/1468-0440.00121
23. Pierce E, Goldstein J (2016) Moving from enterprise risk management to strategic risk management: examining the revised COSO ERM framework. In: 14th Global conference on business and economics. Said School of Business, Oxford, UK
24. Tesfamariam S, Sadiq R, Najjaran H (2010) Decision making under uncertainty—an example for seismic risk management. Risk Anal 30:78–94. https://doi.org/10.1111/j.1539-6924.2009.01331.x
25. Project Management Institute: A guide to the project management body of knowledge
26. Elkington P, Smallman C (2002) Managing project risks: a case study from the utilities sector. Int J Project Manage 20:49–57. https://doi.org/10.1016/S0263-7863(00)00034-X
27. Marle F, Vidal L-A (2011) Project risk management processes: improving coordination using a clustering approach. Res Eng Des 22:189–206. https://doi.org/10.1007/s00163-011-0106-9
28. Aloini D, Dulmin R, Mininno V (2012) Modelling and assessing ERP project risks: a Petri Net approach. Eur J Oper Res 220:484–495. https://doi.org/10.1016/J.EJOR.2012.01.062
29. Zou PXW, Zhang G, Wang J (2007) Understanding the key risks in construction projects in China. Int J Project Manage 25:601–614. https://doi.org/10.1016/J.IJPROMAN.2007.03.001
30. Raz T, Michael E (2001) Use and benefits of tools for project risk management. Int J Project Manage 19:9–17. https://doi.org/10.1016/S0263-7863(99)00036-8
31. Bush JK, Dai WS, Dieck GS, Hostelley LS, Hassall T (2005) The art and science of risk management. Drug Saf 28:1–18. https://doi.org/10.2165/00002018-200528010-00001
32. Keizer JA, Halman JIM, Song M (2002) From experience: applying the risk diagnosing methodology. J Prod Innov Manage 19:213–232. https://doi.org/10.1016/S0737-6782(02)00138-8
33. Grant KP, Cashman WM, Christensen DS (2006) Delivering projects on time. Res Technol Manage 49:52–58. https://doi.org/10.1080/08956308.2006.11657408
34. Boateng P, Chen Z, Ogunlana SO (2015) An Analytical network process model for risks prioritisation in megaprojects. Int J Project Manage 33:1795–1811. https://doi.org/10.1016/J.IJPROMAN.2015.08.007
35. Turner R, Ledwith A, Kelly J (2010) Project management in small to medium-sized enterprises: matching processes to the nature of the firm. Int J Project Manage 28:744–755. https://doi.org/10.1016/J.IJPROMAN.2010.06.005

36. Dandage RV, Mantha SS, Rane SB, Bhoola V (2018) Analysis of interactions among barriers in project risk management. J Industr Eng Int 14:153–169. https://doi.org/10.1007/s40092-017-0215-9
37. Hua J, Shaw R (2020) Corona Virus (COVID-19) “Infodemic” and emerging issues through a data lens: the case of China. Int J Environ Res Public Health 17:2309. https://doi.org/10.3390/ijerph17072309
38. Chatterjee R, Bajwa S, Dwivedi D, Kanji R, Ahammed M, Shaw R (2020) COVID-19 risk assessment tool: dual application of risk communication and risk governance. Progr Disast Sci 7:100109. https://doi.org/10.1016/j.pdisas.2020.100109
39. Mańkowski C, Szmeter-Jarosz A, Jezierski A (2022) Managing supply chains during the Covid-19 pandemic. Central Eur Manage J 30. https://doi.org/10.7206/cemj.2658-0845.91
40. Díaz Pacheco RA, Benedito E (2023) Supply chain response during the COVID-19 pandemic: a multiple-case study. Processes 11:1218. https://doi.org/10.3390/pr11041218
41. Azyabi N (2021) How do information technology and knowledge management affect SMEs’ responsiveness to the coronavirus crisis? Bus Inform 15:75–90. https://doi.org/10.17323/2587-814X.2021.2.75.90
42. Stephens AR, Kang M, Robb CA (2022) Linking supply chain disruption orientation to supply chain resilience and market performance with the stimulus–organism–response model. J Risk Fin Manage 15:227. https://doi.org/10.3390/jrfm15050227
43. Klöckner M, Schmidt CG, Wagner SM, Swink M (2023) Firms’ responses to the COVID-19 pandemic. J Bus Res 158:113664. https://doi.org/10.1016/j.jbusres.2023.113664
44. Margherita A, Heikkilä M (2021) Business continuity in the COVID-19 emergency: a framework of actions undertaken by world-leading companies. Bus Horiz 64:683–695. https://doi.org/10.1016/J.BUSHOR.2021.02.020
45. Zhang J, Mitchell C, Kushniruk A, Guitouni A (2022) Facing disruption: learning from the healthcare supply chain responses in British Columbia during the COVID-19 pandemic. Healthc Manage Forum 35:80–85. https://doi.org/10.1177/08404704211058968
46. van Remko H (2020) Research opportunities for a more resilient post-COVID-19 supply chain—closing the gap between research findings and industry practice. Int J Oper Prod Manage 40:341–355. https://doi.org/10.1108/IJOPM-03-2020-0165
47. Cowling M, Brown R, Rocha A (2020) Did you save some cash for a rainy COVID-19 day? The crisis and SMEs. Int Small Bus J: Research Entrepreneurship 38:593–604. https://doi.org/10.1177/0266242620945102
48. Kuckertz A, Brändle L, Gaudig A, Hinderer S, Morales Reyes CA, Prochotta A, Steinbrink KM, Berger ESC (2020) Startups in times of crisis—a rapid response to the COVID-19 pandemic. J Bus Ventur Insights 13:e00169. https://doi.org/10.1016/J.JBVI.2020.E00169
49. Taleb NN (2008) The Black Swan: the impact of the highly improbable. Penguin Books Ltd
50. Hohenstein N-O (2022) Supply chain risk management in the COVID-19 pandemic: strategies and empirical lessons for improving global logistics service providers’ performance. Int J Logis Manage 33:1336–1365. https://doi.org/10.1108/IJLM-02-2021-0109
51. Corvello V, De Carolis M, Verteramo S, Steiber A (2022) The digital transformation of entrepreneurial work. Int J Entrepreneurial Behav Res 28:1167–1183. https://doi.org/10.1108/IJEBR-01-2021-0067
52. Giotopoulos I, Kontolaimou A, Tsakanikas A (2022) Digital responses of SMEs to the COVID-19 crisis. Int J Entrepreneurial Behav Res 28:1751–1772. https://doi.org/10.1108/IJEBR-11-2021-0924
53. Matt DT, Pedrini G, Bonfanti A, Orzes G (2023) Industrial digitalization. A systematic literature review and research agenda. Eur Manage J 41:47–78. https://doi.org/10.1016/j.emj.2022.01.001
54. Legner C, Eymann T, Hess T, Matt C, Böhmann T, Drews P, Mädche A, Urbach N, Ahlemann F (2017) Digitalization: opportunity and challenge for the business and information systems engineering community. Bus Inf Syst Eng 59:301–308. https://doi.org/10.1007/s12599-017-0484-2

55. Gray J, Rumpe B (2017) Models for the digital transformation. Softw Syst Model 16:307–308. https://doi.org/10.1007/s10270-017-0596-7
56. Alsufyani N, Gill AQ (2022) Digitalisation performance assessment: a systematic review. Technol Soc 68:101894. https://doi.org/10.1016/J.TECHSOC.2022.101894
57. Hein-Pensel F, Winkler H, Brückner A, Wölke M, Jabs I, Mayan IJ, Kirschenbaum A, Friedrich J, Zinke-Wehlmann C (2023) Maturity assessment for Industry 5.0: a review of existing maturity models. J Manuf Syst 66:200–210. https://doi.org/10.1016/j.jmsy.2022.12.009
58. Thordsen T, Murawski M, Bick M (2020) How to measure digitalization? A critical evaluation of digital maturity models. In: Hattingh M, Matthee M, Smuts H, Pappas I, Dwivedi Y, Mäntymäki M (eds) Responsible design, implementation and use of information and communication technology. I3E 2020. Lecture notes in computer science. Springer, pp 358–369. https://doi.org/10.1007/978-3-030-44999-5_30
59. Werner-Lewandowska K, Kosacka-Olejnik M (2019) Logistics 4.0 maturity in service industry: empirical research results. Procedia Manuf 38:1058–1065. https://doi.org/10.1016/j.promfg.2020.01.192
60. Tubis AA (2023) Digital maturity assessment model for the organizational and process dimensions. Sustainability 15:15122. https://doi.org/10.3390/su152015122
61. Rojko A (2017) Industry 4.0 concept: background and overview. Int J Interact Mob Technol (iJIM) 11:77. https://doi.org/10.3991/ijim.v11i5.7072
62. Hossain MR, Akhter F, Sultana MM (2022) SMEs in Covid-19 crisis and combating strategies: a systematic literature review (SLR) and a case from emerging economy. Oper Res Perspec 9:100222. https://doi.org/10.1016/J.ORP.2022.100222
63. https://www.mckinsey.com/capabilities/strategy-and-corporate-finance/our-insights/how-covid-19-has-pushed-companies-over-the-technology-tipping-point-and-transformed-business-forever. How COVID-19 has pushed companies over the technology tipping point—and transformed business forever (2020)
64. Almeida F, Duarte Santos J, Augusto Monteiro J (2020) The challenges and opportunities in the digitalization of companies in a post-COVID-19 world. IEEE Eng Manage Rev 48:97–103. https://doi.org/10.1109/EMR.2020.3013206
65. Amankwah-Amoah J, Khan Z, Wood G, Knight G (2021) COVID-19 and digitalization: the great acceleration. J Bus Res 136:602–611. https://doi.org/10.1016/j.jbusres.2021.08.011
66. Renault BY, Agumba JN, Balogun OA (2016) Drivers for and obstacles to enterprise risk management in construction firms: a literature review. Procedia Eng 164:402–408. https://doi.org/10.1016/j.proeng.2016.11.637
67. Liebenberg AP, Hoyt RE (2003) The determinants of enterprise risk management: evidence from the appointment of chief risk officers. Risk Manage Insurance Rev 6:37–52. https://doi.org/10.1111/1098-1616.00019
68. Kang S, Kim E, Shim J, Cho S, Chang W, Kim J (2017) Mining the relationship between production and customer service data for failure analysis of industrial products. Comput Ind Eng. 106:137–146. https://doi.org/10.1016/J.CIE.2017.01.028
69. Haksöz Ç (2013) Risk intelligent supply chains: how leading Turkish companies thrive in the age of fragility. Taylor and Francis Group/CRC Press, Boca Raton
70. Wu DD, Chen SH, Olson DL (2014) Business intelligence in risk management: some recent progresses. Inf Sci (NY) 256:1–7. https://doi.org/10.1016/J.INS.2013.10.008
71. Er Kara M, Oktay Fırat SÜ, Ghadge A (2020) A data mining-based framework for supply chain risk management. Comput Ind Eng 139:105570. https://doi.org/10.1016/j.cie.2018.12.017
72. Apgar D (2006) Risk intelligence: learning to manage what we don't know. Harvard Business Press
73. Wu L, Yue X, Jin A, Yen DC (2016) Smart supply chain management: a review and implications for future research. Int J Logist Manage 27:395–417. https://doi.org/10.1108/IJLM-02-2014-0035
74. Liu L, Daniels H, Hofman W (2014) Business intelligence for improving supply chain risk management. In: Hammoudi S, Cordeiro J, Maciaszek L, Filipe J (eds) Enterprise information systems. ICEIS 2013. Lecture notes in business information processing. Springer, pp 190–205. https://doi.org/10.1007/978-3-319-09492-2_12

75. Lee E-B, Kim J, Lee S-G (2017) Predicting customer churn in mobile industry using data mining technology. Industr Manage Data Syst 117:90–109. https://doi.org/10.1108/IMDS-12-2015-0509
76. Ranjan J, Bhatnagar V (2011) Role of knowledge management and analytical CRM in business: data mining based framework. Learn Org 18:131–148. https://doi.org/10.1108/09696471111103731
77. Witten I, Frank E, Hall MA, Pal CJ (2017) Data mining: practical machine learning tools and techniques. Morgan Kaufmann
78. Larose DT, Larose CD (2015) Data mining and predictive analytics. John Wiley & Sons
79. Kim Y, Chen Y, Linderman K (2015) Supply network disruption and resilience: a network structural perspective. J Oper Manage 33–34:43–59. https://doi.org/10.1016/j.jom.2014.10.006
80. Kamalahmadi M, Parast MM (2016) A review of the literature on the principles of enterprise and supply chain resilience: major findings and directions for future research. Int J Prod Econ. 171:116–133. https://doi.org/10.1016/j.ijpe.2015.10.023
81. Priya Datta P, Christopher M, Allen P (2007) Agent-based modelling of complex production/distribution systems to improve resilience. Int J Logist Res Appl 10:187–203. https://doi.org/10.1080/13675560701467144
82. Christopher M, Peck H (2004) Building the resilient supply chain. Int J Logist Manage 15:1–14. https://doi.org/10.1108/09574090410700275
83. Ponomarov SY, Holcomb MC (2009) Understanding the concept of supply chain resilience. Int J Logis Manage 20:124–143. https://doi.org/10.1108/09574090910954873
84. Johnson N, Elliott D, Drake P (2013) Exploring the role of social capital in facilitating supply chain resilience. Supply Chain Manage: Int J 18:324–336. https://doi.org/10.1108/SCM-06-2012-0203
85. Roberta Pereira C, Christopher M, Lago Da Silva A (2014) Achieving supply chain resilience: the role of procurement. Supp Chain Manage: Int J 19:626–642. https://doi.org/10.1108/SCM-09-2013-0346
86. Ponis ST, Koronis E (2012) Supply chain resilience: definition of concept and its formative elements. J Appl Bus Res (JABR) 28:921. https://doi.org/10.19030/jabr.v28i5.7234
87. Craighead CW, Blackhurst J, Rungtusanatham MJ, Handfield RB (2007) The severity of supply chain disruptions: design characteristics and mitigation capabilities. Decis Sci 38:131–156. https://doi.org/10.1111/j.1540-5915.2007.00151.x
88. Wu T, Blackhurst J, O'grady P (2007) Methodology for supply chain disruption analysis. Int J Prod Res 45:1665–1682. https://doi.org/10.1080/00207540500362138
89. Chopra S, Sodhi MS (2004) Managing risk to avoid supply chain breakdown. MIT Sloan Manag Rev 46:53–62
90. Aqlan F, Lam SS (2016) Supply chain optimization under risk and uncertainty: a case study for high-end server manufacturing. Comput Ind Eng 93:78–87. https://doi.org/10.1016/J.CIE.2015.12.025
91. Brusset X, Teller C (2017) Supply chain capabilities, risks, and resilience. Int J Prod Econ 184:59–68. https://doi.org/10.1016/J.IJPE.2016.09.008
92. Rajagopal V, Prasanna Venkatesan S, Goh M (2017) Decision-making models for supply chain risk mitigation: a review. Comput Ind Eng. 113:646–682. https://doi.org/10.1016/j.cie.2017.09.043
93. Hudnurkar M, Deshpande S, Rathod U, Jakhar SK (2017) Supply chain risk classification schemes: a literature review. Oper Supp Chain Manage: Int J 10:182–199. https://doi.org/10.31387/oscm0290190
94. Christopher M, Mena C, Khan O, Yurt O (2011) Approaches to managing global sourcing risk. Supp Chain Manage: Int J 16:67–81. https://doi.org/10.1108/13598541111115338
95. Soni U, Jain V, Kumar S (2014) Measuring supply chain resilience using a deterministic modeling approach. Comput Ind Eng 74:11–25. https://doi.org/10.1016/J.CIE.2014.04.019
96. Waters D (2007) Supply chain risk management: vulnerability and resilience in logistics. Kogan Page

97. Mandal S (2012) An empirical investigation into supply chain resilience. IUP J Supp Chain Manage IX:46–61
98. Tubis AA, Poturaj H, Smok A (2024) Interaction between a human and an AGV system in a shared workspace—a literature review identifying research areas. Sustainability 16:974. https://doi.org/10.3390/su16030974
99. Matthyssens P (2019) Reconceptualizing value innovation for Industry 4.0 and the Industrial Internet of Things. J Bus Industr Mark 34:1203–1209. https://doi.org/10.1108/JBIM-11-2018-0348
100. Sony M, Naik S (2020) Industry 4.0 integration with socio-technical systems theory: a systematic review and proposed theoretical model. Technol Soc 61:101248. https://doi.org/10.1016/J.TECHSOC.2020.101248
101. Howard RA (2007) The foundations of decision analysis revisited. In: Edwards WM, von Winterfeldt D (eds) Advances in decision analysis. Cambridge University Press, pp 32–56. https://doi.org/10.1017/CBO9780511611308.004

Chapter 2
Risk in Anthropotechnical Systems

Contents

Chapter 1 presents the fundamental concepts of risk and managing it. Many authors point out that even if individual teams within an organisation understand the concept of risk and its management, their understanding and risk assessment will vary significantly between different parts of the organisation and the type of organisation [1]. Managers in finance will hold a different approach to risk, where the risks assessed can have positive and negative impacts, and the analyses are primarily quantitative. A markedly different perception of risk will be held by managers in the supply chain area, where risk has a mainly negative and often unquantifiable dimension that will compel the use of qualitative rather than purely quantitative analyses. For this reason, at the operational level, it makes sense to adapt risk assessment methods to the specific functioning of the area under analysis, thus considering the conditions of the processes and systems supporting the operations performed. In responding to these needs, however, it is necessary to comprehensively recognise the various approaches to risk management, which will help to formulate assumptions for the model to be developed.

Research conducted by the Author [2–6] indicates that developing a new approach to risk management for logistics processes requires an interdisciplinary view of the analysed research aspects related to implementing selected operational processes. Above all, analyses of the risks of internal transport operations in a given enterprise require the integration of two analytical trends, namely business and engineering

A. Tubis, *The New Paradigm of Risk in Internal Transport Supporting Logistics 4.0 System*, Lecture Notes in Intelligent Transportation and Infrastructure,
https://doi.org/10.1007/978-3-031-64920-2_2

approaches. Therefore, it is necessary to formulate new guidelines for risk assessment based on the concepts of business process management and operational risk in the enterprise while considering the safety and reliability of technical systems supporting the realisation of logistics processes. The answer to this need is the concept of a holistic risk assessment system for logistics processes, which integrates the needs indicated above. The framework for this concept is formulated based on the assumptions of the four organisational risk management strategies outlined in Sects. 2.1 and 2.2.

2.1 The Concept of Risk in the Business Approach

According to the definitions given in Chap. 1, risk in business terms refers primarily to the impact of uncertainty on the achievement of objectives defined for the organisation. However, this approach does not indicate the manner (direction) of impact on goal achievement, i.e. risk refers to events that can affect goal achievement in both positive and negative ways. In their research, Bromiley et al. [1] highlight that papers published in the '90s already emphasise that risk management should go beyond the traditional goal of exclusive "variance minimisation", focusing on the negative sides of risk. Stulz [7] even states that an organisation should reduce risk exposures in those areas of its business where it does not have a comparative advantage while exploiting risk where there is an advantage. This recommendation indicates the dual nature of the concept of risk found in the business approach.

2.1.1 Enterprise Risk Management and Standard ISO 31000

The traditional approach to risk management in organisations was based on the assumption of the need to separate the management of individual risk categories between separate organisational units [8]. This resulted in a high disaggregation of risks, which were managed separately by individual units in the organisation. This approach created so-called risk silos, which were analysed in a disconnected manner without examining the interconnecting relationships and interactions between them. Financial organisations managed credit, market, operational and liquidity separately so that these risks were assessed and managed within individual silos. Non-financial companies also followed a similar approach to managing financial, strategic and operational risks [8]. Based on criticism of this approach to risk, the concept of Enterprise Risk Management (ERM) was proposed, which was defined as "a structured and disciplined approach to help management understand and manage uncertainties and encompasses all business risks using an integrated and holistic approach" [9]. A framework was developed for the concept, articulated in the document "Enterprise Risk Management—Integrated Framework", published by the Committee of Sponsoring Organization of the Treadway Commission in 2004. This document proposed a framework for the ERM concept in the form of a cube presenting eight components critical to the success of ERM implementation, which should relate to the four risk areas that may exist at the four levels of the organisation. The components embedded in the ERM framework include [10]:

- Internal environment, which determines attitudes towards risk, shapes risk appetite among managers and formulates ethical values.
- An objective setting consistent with the identified risk appetite based on correctly realised risks.
- Risk assessment is prepared based on a correctly mapped probability and impact of risk occurrence and identifying inter-relationships between risks.
- Risk response that is appropriate to risk appetite and tolerance.
- Control activities, which ensure the effectiveness of the risk response through implemented policies and procedures.
- Information and communication systems enable data to be identified, collected, and communicated in a format that is tailored to the needs of managers and employees.
- A risk monitoring process that is implemented continuously or periodically so that it can be modified when necessary.

In an updated document entitled "Enterprise Risk Management. Integrating with Strategy and Performance", published in 2017, COSO has reduced the number of components constituting the ERM framework to five items. However, implementation principles have been formulated for each of them. In doing so, COSO emphasises that it is necessary to implement all 20 defined principles to integrate enterprise risk management with strategy and performance effectively [11]. The components and the principles assigned to them are shown in Fig. 2.1.

An interesting comparison of the two frameworks (2004 and 2017) is presented in [12], among others.

ERM is a concept focused on the systematic and integrated management of the total risk faced by a company [13]. Harrington et al. [14] further emphasise that a company should identify, measure (assess) and manage all its exposures to different types of risk (financial, strategic, operational) in a single, integrated system as opposed to the siloed approach used to date. Thus, unlike the traditional approach, ERM recognises that each highlighted risk category is part of an organisation's overall risk portfolio, which is managed holistically [8]. It should also be noted that the traditional approach to risk management focused primarily on protecting the company from adverse financial consequences. In contrast, in the ERM concept, risk management becomes an important strategic element, enabling the organisation to make better decisions considering the different types of risks involved [15]. In the ERM concept, risk management must be linked to the organisation's business strategy, goal setting and managerial decision-making [16]. Therefore, many authors recommend moving away from a fragmented, ad hoc and narrow approach to risk management to a comprehensive and unified system as defined in the ERM concept [1]. In this way, the risk management strategy can support the competitive advantage of the company being built and support its growth [17].

COMPONENT	DESCRIPTION	PRINCIPLES
GOVERNANCE AND CULTURE	Governance is important for effective risk management; it defines the role of risk assessment in decision-making processes and establishes the responsibilities of participants in the process. Culture shapes ethical attitudes, desired behaviours and the way risk is understood within an entity.	**1.** Exercises Board Risk Oversight **2.** Establishes Operating Structures **3.** Defines Desired Culture **4.** Demonstrates Commitment to Core Values **5.** Attracts, Develops and Retains Capable Individuals
STRATEGY AND OBJECTIVE-SETTING	The risk appetite defined for the organisation should be the basis for developing strategy and setting goals for managers to achieve	**6.** Analyses Business Context **7.** Defines Risk Appetite **8.** Evaluates Alternative Strategies **9.** Formulates Business Objectives
PERFORMANCE	The assessed risks are related to the set business objectives and should be prioritised according to the defined risk appetite. A portfolio assessment of the magnitude of the risks taken should be adopted to decide how to respond to risks. The results should be reported to stakeholders	**10.** Identifies Risk **11.** Assesses Severity of Risk **12.** Prioritises Risk **13.** Implements Risk Responses **14.** Develops a Portfolio View
REVIEW AND REVISION	The unit's performance should be subject to a review to assess the effectiveness of the risk management changes implemented and to identify what changes are still needed	**15.** Assesses Substantial Change **16.** Reviews Risk and Performance **17.** Pursues Improvement in ERM
INFORMATION, COMMUNICATION AND REPORTING	ERM requires an efficient and effective reporting and information-sharing system concerning both internal and external sources.	**18.** Leverages Information Systems **19.** Communicates Risk Information **20.** Reports on Risk, Culture and Performance

Fig. 2.1 Components of the 2017 ERM framework (based on [11])

The ERM concept focuses on assessing and responding to the different risk categories. At the same time, these categories relate to business objectives formulated for the organisation, which relate to [18]:

- Strategy—the achievement of the organisation's global objectives that reflect the adopted direction of the organisation.
- Operations—effective and efficient use of resources and asset management.
- Reporting—ensuring an efficient reporting system that provides reliable and timely information to support decision-making processes.
- Compliance—ensuring that the organisation's operations comply with applicable laws and regulations.

Implementing such an integrated and, at the same time, comprehensive approach to risk management in an organisation is a major strategic challenge for managers but also an operational challenge as it affects all members of the organisation. This is because adopting the ERM approach brings about changes in the organisation's culture and forces a change in how decisions and actions are taken. It is the responsibility of the company's management to determine the so-called risk appetite influencing managerial decisions and to develop a risk management policy that will affect the management of the organisation's activities. Arguments in favour of this challenge are the benefits of implementing an ERM approach, which is widely reported in reports and academic publications. According to [19], companies implementing ERM across the organisation can achieve several benefits, most notably:

- Increasing the range of opportunities by identifying new challenges and risks.
- Identify risks in different parts of the organisation and manage them from an enterprise-wide perspective to improve performance.
- Identify risks and select an appropriate response that reduces surprises and associated losses while reaping the benefits of the changes.
- Reducing variability in performance through risk estimation translates into minimising disruption and maximising the exploitation of opportunities.
- Improving resource utilisation by prioritising and improving resource allocation based on risk assessment.
- Increasing the resilience of the business based on better anticipation and response to change.

A study by Renault et al. [20] indicates that the following benefits of ERM implementation are most frequently cited in the literature on ERM:

• reduce costs and losses, • improved decision-making, • increased profitability and earnings, • better risk reporting and communication, • increased management accountability	• greater management consensus, • competitive advantages, • improved owners' satisfaction, • better resource allocation, • improved control of an enterprise on its projects, • reduced earnings volatility

Liebenberg and Hoyt [8] found additional rationale for implementing the ERM concept in the internal and external factors shown in Fig. 2.2.

Identifying the benefits of ERM implementation and promoting them within the organisation is essential for management to gain the necessary support for the ERM programme and reinforce positive outcomes. However, identifying obstacles that may occur during the implementation of this approach remains equally important. Knowing the potential obstacles enables management to identify the implementation challenges clearly and take preventive and corrective actions in advance to reduce the undesirable effects of these disruptions. Based on a literature review, Renault et al.

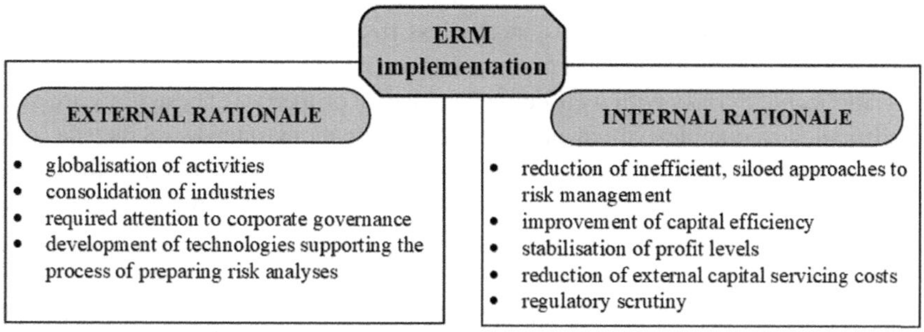

Fig. 2.2 The rationale for implementing the ERM concept (based on [8])

defined the most commonly reported obstacles occurring in ERM implementation [20]:

- insufficient knowledge of managers regarding risk management in the organisation;
- compromising attitudes of managers hindering risk discussions;
- management priorities not promoting risk management issues;
- reluctance to discuss sensitive information across business units;
- uncertainty about how to add value to the company by implementing ERM;
- the complex nature of risk management, which requires specialist knowledge, skills and experience relating to specific areas of the business;
- the lack of alignment of the risk management strategy with the organisation's overall business strategy;
- the lack of integration of risk management into the organisation's processes;
- the lack of a common risk language, which is required in discussions to fully understand the phenomena occurring and their impact on the organisation's operations;
- the high level of difficulty in quantifying risk levels and the lack of understanding by managers of simple assessment tools;
- Lack of data required to perform risk assessments: (a) inadequate data quality; (b) limited access due to insufficient integration between systems; (c) lack of risk mapping and modelling tools;
- segmented (siloed) approach to the different types of risks in place in the company.

Alijoyo et al. [21] emphasise that risk management is different for each organisation due to their operations' internal and external contexts. Indeed, the risk is not universal and varies depending on the enterprise's sector but also on its business processes, the level of digitisation, the products produced and the target market of customers. Therefore, when implementing the ERM concept, each organisation has its own perspective, priorities, and needs when dealing with risks. Of course, the primary objectives and principles recommended by ERM are the same, but the degree and nature of the risks companies face are quite different [21].

Bromiley et al. [1] found in 2015 that academic research on ERM appeared mainly in accounting and finance journals and far less frequently in management journals. However, it is worth noting that finance and accounting research focuses on tools only applicable to risks with well-defined statistical properties. These tools are also mathematically complex, often have limited applications outside of finance, and may not be understood by managers. Today, such a narrow perception of the scope of application of ERM concepts has changed. A literature review by Anton and Nucu [22] indicates that although the most significant number of papers, the most cited, appear in journals such as the Journal of Risk and Insurance, Accounting Organisations and Society, and Contemporary Accounting Research, equally popular in recent years are papers placed in journals such as International Journal of Production Research, Journal of Construction Engineering and Management and Production Planning Control (citation index h > 30). Based on a review of 101 papers, Anton

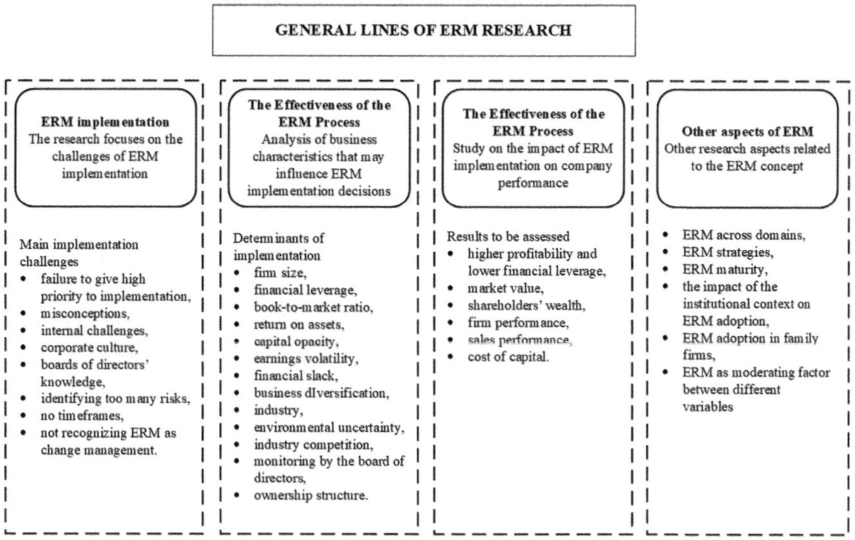

Fig. 2.3 Research areas on the ERM concept (based on [22])

and Nucu [22] also identified four main research areas on the ERM concept, shown in Fig. 2.3.

Bromiley et al. [1] also argue that there are many ambiguities and misunderstandings in ERM research, which can make it difficult to fully understand the conceptual assumptions and hinder the implementation process in an organisation. For this reason, these authors have formulated the basic assumptions of the ERM concept [1]:

- Managing the risk of the entire enterprise portfolio is more effective than managing the risk of each area (department).
- Every decision made by managers in an enterprise should be linked to the risk management process. Even a single decision can be linked to multiple risks at different stages of its implementation.
- ERM encompasses traditional risks (product liability and accidents) and strategic risks (product obsolescence, competitive actions). Often, the most significant risks lie in strategic areas where there is a lack of historical data to estimate the likelihood of their occurrence.
- ERM assumes that companies should not view risks as a problem to be reduced or eliminated. If a company can manage a particular risk, it should build a competitive advantage.

Based on the ERM concept, the ISO 31000 standard, first published in 2009 and updated in 2018, was developed. According to the standard, risk management should be defined as the coordinated activities to guide and control an organisation concerning the emerging risks associated with its activities [23]. The standard describes a risk management architecture comprising principles, structure, and process. At the same time, Alijoyo and Norimarna point out the relationships between the distinguished elements of the architecture of ISO 31000 Standard [24]:

- The indicated core elements are interdependent, and their implementation cannot be disconnected.
- The defined risk management principles are critical to a successful implementation process and must be considered in the other two elements.
- The framework developed formulates the critical findings that, if implemented effectively at all levels of the organisation, can ensure the success of risk management.
- The risk management process should be an integral part of the business processes and corporate culture and should be tailored to the needs and specifics of the company.

The risk management process recommended in ISO 31000 is presented as a procedure in Fig. 1.1. For the procedure so defined, ISO 31000:2018 [23] also defines eight risk management principles that should be followed to effectively and efficiently achieve the objectives defined for the organisation. These principles are shown in Fig. 2.4.

Research confirms that applying the methodology recommended in the standard is conducive to reducing financial and operational losses, reducing inventory losses, reducing reliance on external capital and improving supplier relationships [25]. Of course, the organisation's maturity level in implementing the risk standards defined in ISO 31000 is an important issue. Various models for assessing risk management maturity can be found in the literature [26], including a model referring exclusively to ISO 31000[23].

In conclusion, it should be noted that many authors consider the framework proposed by COSO and recommended in ISO 31000 as a highly abstract conceptual framework that does not allow the formulation of control objectives at the required level of detail [27]. In addition, many popular standards describe general phases of the risk management process, often omitting detailed aspects related to process management and the formulation of the required procedures [12]. Even such important terms as risk culture or risk appetite are not defined clearly and unambiguously. At the same time, all frameworks are characterised by a similar approach and distinguish comparable activities about the risk management process. Thus, based on the two frameworks presented above and the AS/NZS 4360 standard, Rubino proposed a consensus framework for implementing the risk management process, which includes the four stages of the procedure shown in Table 2.1.

The company's objectives are strongly correlated with the environment represented by customers, suppliers, the government and other organisations. This environment strongly influences the enterprise's core business (e.g., the customer and its needs shaping its offer) and the supporting processes (suppliers' competitiveness, business regulations). Therefore, this environment thus affects the company's

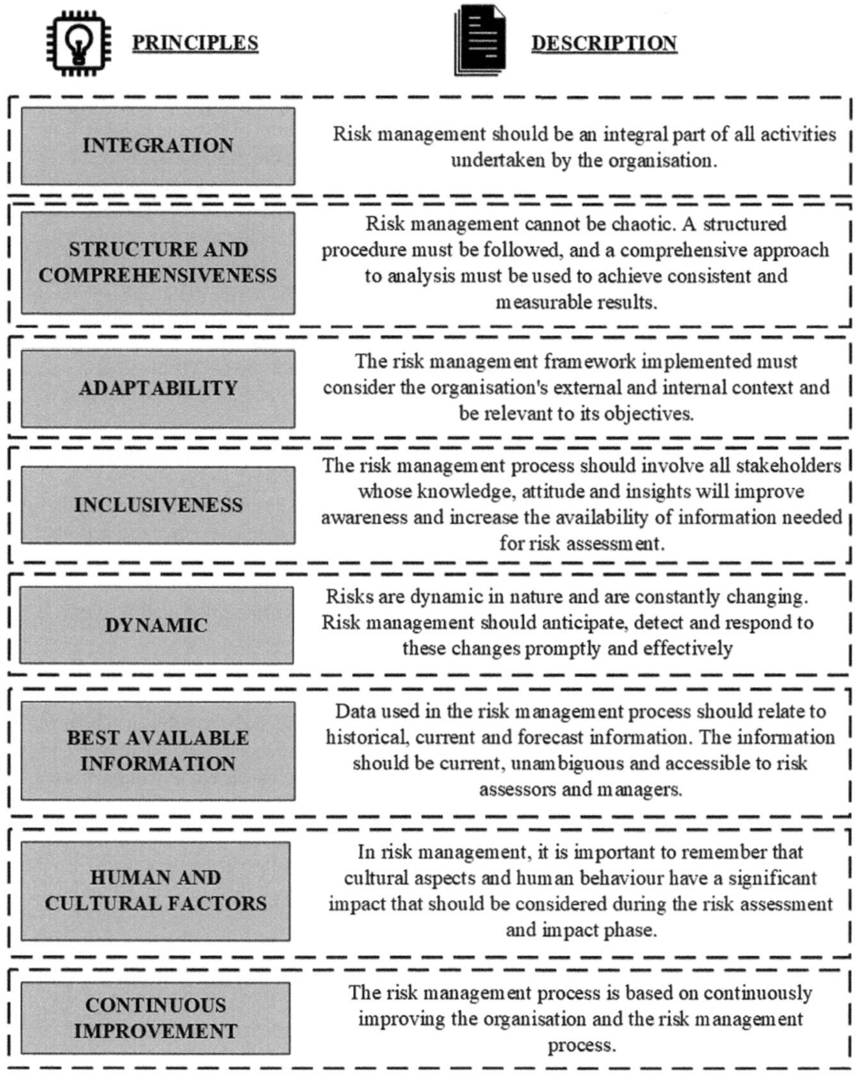

Fig. 2.4 Risk management principles (based on [23])

objectives and becomes a source of potential risk [25]. Soltanizadeh et al. [18] note, however, that the ERM concept focuses on risks assessed from the point of view of a particular organisation. Therefore, risks relating to relationships with suppliers, distributors, and customers are analysed only in the context of the business environment. In this approach, the external risks identified by the company consist of the risks generated by other actors to the company and the risks faced by the company's relationships with other actors in the business environment [25]. Therefore, this approach lacks a broader view of risk assessment that looks beyond the individual

Table 2.1 Stages of implementing a risk management process (based on [12])

Stage name	Stage characteristics
Stage 1: Understanding the organisation and its internal and external context	• The organisational environment—the philosophy of the top management in the risk and its levels of acceptability, the culture of risk and behaviour of people operating at all levels of the company, the managerial style, the integrity and ethical values, skills, definition of areas of authority and responsibility, the existence of appropriate policies and procedures; • The external context—the social and cultural, political, legal, regulatory, financial, technological, economic, natural and competitive environment, whether international, national, regional or local; key drivers and trends having an impact on the objectives of the organisation; and relationships with, perceptions and values of external stakeholders; • The objectives setting—establishing the organisation's vision, goals, principles, resources required, responsibilities and authorities, and the records to be kept
Stage 2: Risk management activity	• There is a need to identify internal and external events that positively or negatively affect the organisation's objectives • Conducting a risk assessment of the identified events—depending on the likelihood and impact of the assessed events, managers prioritise actions • Responses to the occurring risks are assessed for the prioritised events, and actions are taken to align the identified risks with the risk tolerance and appetite. The risk response may include one or more risk modification options
Stage 3: Control activities and monitoring	• Develop, in the form of procedures, controls to mitigate operational risks and identify actual risks • Control activities should make it possible to identify possible events or risk factors; implement changes in policies and procedures and adopt new control procedures; monitor the effectiveness of the controls already implemented; continuously monitor the effectiveness of the functioning of the risk management process
Stage 4: Information and communication	• Continuous communication with external and internal stakeholders • Frequent reporting of results. The information system reports internal data related to events and the external situation • Identifying, collecting and distributing information in the proper form and at the right time so that all units can fulfil their responsibilities • High-quality communication processes that depend on the quality of the information system

enterprise and includes the perspective of a group of organisations operating within the chosen supply chain. Therefore, when discussing the business approach to risk, it is necessary to change the analytical perspective and elevate it to the level of supply chain management. This perspective is presented in the concept of supply chain risk management (SCRM). This concept was identified in Chap. 1 as one of the pillars of the current trend towards supply chain resilience. From the point of view of the subject matter of this book, SCRM takes on particular importance because supply chain risk consists, among other things, of risks occurring in individual logistics systems, including internal logistics service processes.

However, before the SCRM concept is presented, it is worth considering the idea of supply risk itself. In publications at the beginning of the twenty-first century, this concept was equated by many authors with risks concerning the Buyer–supplier relationship [28, 29]. Zsidisin looked at the issue more broadly and defined supply riks as "the probability of an incident associated with inbound supply from an individual supplier failure or the supply market occurring, in which its outcomes result in the inability of the purchasing firm to meet customer demand or causes threats to customer life and safety [30]". This dichotomy was later developed by Peck adding that it is "anything that disrupts or impedes the information, material or product flows from original suppliers to the delivery of the final product to the ultimate end user [31]". However, Ho et al. [32] pointed out that all the definitions of supply risk focus on a selected function or part of the supply chain and do not cover the entire flow. Therefore, these authors defined supply risk as "the likelihood and impact of unexpected macro and/or micro level events or conditions that adversely influence any part of a supply chain leading to operational, tactical, or strategic level failures or irregularities [32]".

Supply chain risk is not a new phenomenon. Companies have constantly been exposed to the possibility that a supplier will fail to deliver within the required timeframe, will deliver against quality requirements or in a different quantity than expected, or will adjust the price at the last minute. However, the new approach to supply chain risk is characterised by two features:

- Firstly, even if the risk of a particular event for a single member of the supply chain is small, the cumulative effect for all members already becomes significant.
- Secondly, in developing ways of mitigating the risks involved, reference can be made not only to the resources of an individual organisation but also to the potential of the entire supply chain.

Supply chain risk management is defined in many ways in the literature. Some definitions focus on the risk management process [32, 33], often described as 4-stages: identification, assessment, treatment and risk monitoring. Other authors primarily focus their definitions on the objectives set for SCRM [34, 35]. Among these objectives, the following are most commonly distinguished [36]: ensure profitability, cost saving, reduce vulnerability, and continuity. Based on these different approaches, Fan et al. formulated a proposal for a comprehensive definition of the term SCRM, including "the identification, assessment, treatment, and monitoring of supply chain risks, with the aid of the internal implementation of tools, techniques, and strategies and external coordination and collaboration with supply chain members to reduce vulnerability and ensure continuity coupled with profitability, leading to competitive advantage [36]". What is central to developing the SCRM concept is the need for organisations to understand that their vulnerability to risk depends mainly on the other supply chain participants [37].

Supply chain risk analysis aims to identify the most significant adverse events that may affect the execution of critical business processes, such as product and service development, production control, customer and supplier relationship management, and sales and operations planning [25]. Therefore, the first stage of risk assessment related to identifying potential adverse events requires special attention and knowledge not only of individual organisations but also of the relationships between them

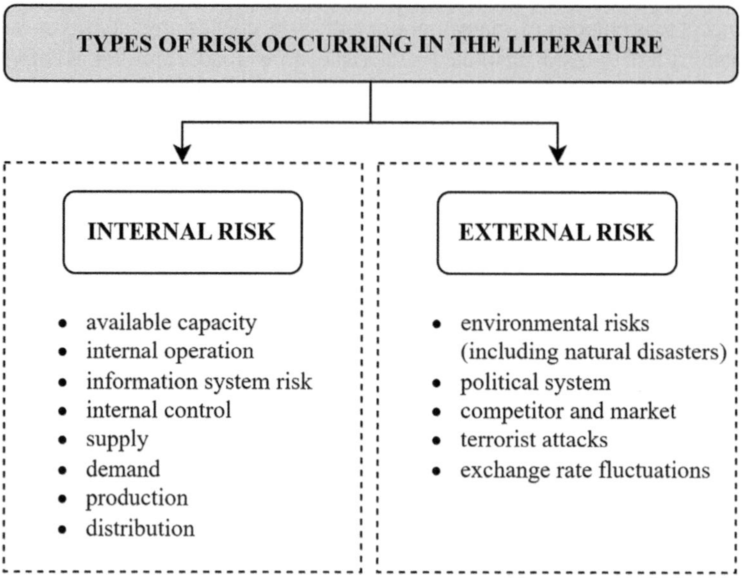

Fig. 2.5 Two types of risk occurrence (based on [32])

affecting the functioning of the entire supply chain. A literature review by Ho et al. [32] indicates that most study authors identify two types of risk occurrence, as shown in Fig. 2.5.

It is worth noting, however, that some authors dealing with SCRM recognise the division into three groups:

- Internal risk includes events related to disruptions occurring at the level of the organisation;
- External risk—events related to the environment;
- The third group consists of risks related to the functioning of the supply chain, the so-called network-related risks. These risks are external from the point of view of the supply chain participant but internal from the point of view of the supply chain network. Christopher and Peck [38] include supply and demand risks in this group.

Based on their literature review, Ho et al. [32] proposed a classification of risk factors into macro and micro factors. They then identified the most commonly reported risk factors in the literature, ranking them according to the proposed classification. The framework of the proposed classification and examples of risk factors is shown in Fig. 2.6.

Qualitative and quantitative methods are used to assess risk, and we distinguish between analytical and empirical methods. Qualitative risk assessment methods estimate the loss incurred based on the knowledge and judgement of the expert responsible for preparing the risk analysis rather than precise financial values, whereas

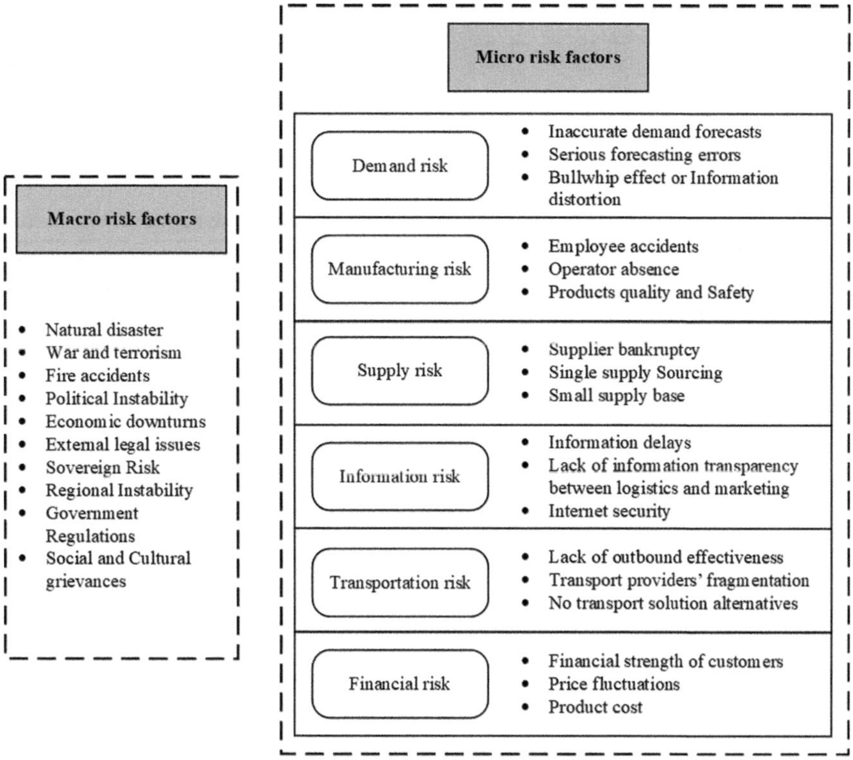

Fig. 2.6 Classification of risk factors (based on [32])

quantitative methods of analysis measure risk based on a monetary or discrete value [39]. In doing so, Kendrick's research [40] demonstrates that quantitative methods are more precise and provide more information about each identified risk. It is also worth noting that quantitative risk analyses are now strongly supported by numerous computer software models. The research presented in [32] also confirms that quantitative analytical methods are becoming increasingly popular, and their use is growing yearly. Among the most popular methods in this group are mathematical programming, followed by newsvendor model and simulation. More interest needs to be placed on the use of quantitative empirical methods by researchers. However, as Ho et al. note, this may be primarily due to the limitations in the collaboration between researchers and industry, which prevent the implementation of empirical studies and access to the required data [32]. However, it is worth noting that the most popular method in this group is the multiple regression model.

Fan et al. [36] note that for the risk assessment process to deliver the expected results at the supply chain level, it is necessary to incorporate the factors shown in Fig. 2.7 into the analysis.

Concerning supply chain risk, it should be noted that some risks involve more interrelationships than others, so their impact strength is definitely higher. This is

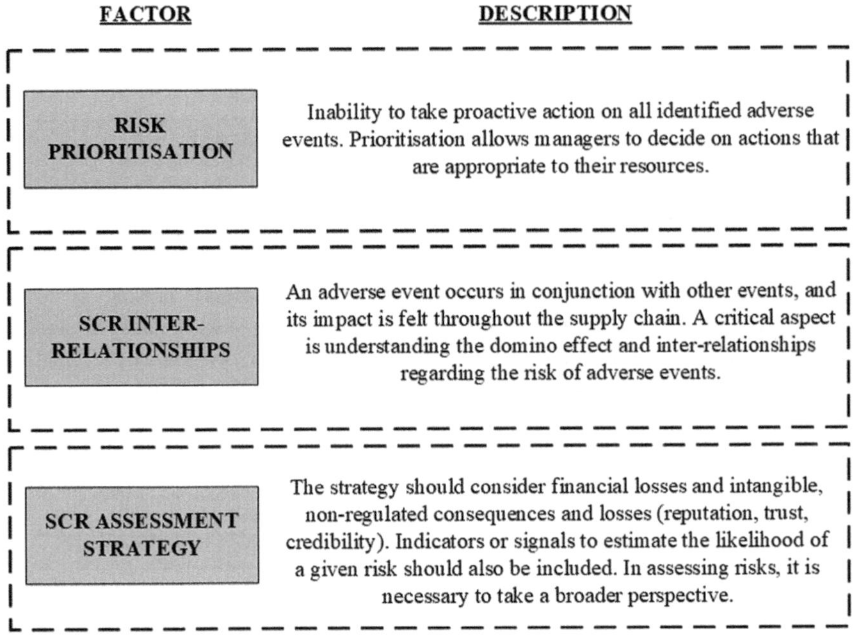

Fig. 2.7 Factors assessed in the risk assessment process at supply chain level (based on [36])

why some authors (e.g. [41]) focus on identifying critical risks, which can trigger many other events and have a knock-on effect. Sarker et al., in their study [42], proposed to divide the types of dependencies between different kinds of risks into (a) a positive dependency, meaning that the removal of one risk mitigates the effects of one or more other events); and (b) a negative dependency meaning that the removal of one risk generates one or more other.

Much of the research on the SCRM concept focuses primarily on establishing the relationship between disruptions occurring and performance [43, 44]. In contrast, a second critical area of research is selecting an appropriate risk mitigation strategy and its importance in enhancing supply chain performance. Talluri et al. [45] even emphasise that the selection of risk mitigation is a critical component of a risk management strategy. Analysing various factors influencing buyer–supplier relationships, Bode et al. [46] found that trust and dependence between partners strongly influence mitigating supply chain disruptions. Mishra et al. [44] developed a theoretical model to explain how to select an appropriate strategy to reduce the risk in the buyer–supplier relationship and improve supply chain performance. Bode et al. [46] also highlighted in their study that selecting a firm's response to supply chain disruptions is based on information processing and resource-based theory. Talluri et al. [45] argued that the adequacy and effectiveness of the selection of risk mitigation strategies depends on the conditions present in the external and internal environment. Considering these arguments, many researchers distinguish two basic risk mitigation strategies:

(1) Buffering is the most commonly used risk mitigation strategy by companies. This strategy protects the supply chain from disruptions by maintaining adequate backup buffers. Chattopadhyay et al. [47] recommended a buffering strategy as a hedge to minimise the risks associated with supply by the incumbent supplier in their study. According to Sinha et al. [48], this strategy should be aimed at identifying where stocks should be held and the appropriate level of stocks to ensure continuity of supply. According to this approach, the buffering strategy acts as a shield for the organisation, protecting it from disruptions occurring at different levels and locations in the supply chain. However, this implies generating free resources to be activated when a disruption occurs. Some authors point out in this regard that a buffering strategy should respond to the risks associated with the occurrence of events that cannot be eliminated by process improvement [44].
(2) Bridging strategy—protects the organisation from the risk of adverse events by establishing strong relationships with exchange partners. It involves developing relationships with relevant partners up to full integration. In their research, Bode et al. [46] argued that this is an internal approach to managing uncertainty by engaging in activities related to crossing or pushing boundaries with supply chain partners. These activities may involve creating a common risk management system or integrating and improving information systems. For this reason, Premkumar et al. [49] analysed the concept of building bridges between partners through the lens of information theory. They found that easy access to authentic and timely information regarding the occurrence of supply chain disruptions is a way to process the way associated risks are managed.

Based on a review of 126 publications on SCRM, Rajagopal et al. [50] distinguished five risk mitigation strategies, shown in Fig. 2.8. A critical analysis of the research on all strategies allowed the authors to conclude that robust/resilient SCND and RPA are the most commonly used and described risk mitigation strategies. Second in terms of popularity is the reliable facility location/fortification and inventory management strategy followed by SSOA.

The level of trust and dependency between partners is crucial in selecting an appropriate mitigation strategy. Based on these, the strategy that will bring the most significant benefit to the company in terms of minimising risk and improving the results obtained is selected. Wang et al. [51] argue in their research that trust plays a critical role in risk mitigation by reducing uncertainty arising from partner actions and opportunistic behaviour. At the same time, Yeung et al. [52] demonstrate that trust additionally facilitates overcoming social difficulties and conflicts and fosters the strengthening of long-term relationships between supply chain links. The interdependence of partners is also an essential part of business relationships in supply chains that help to achieve desired outcomes. Handley and Benton [53] argue that dependencies as an inter-organisational phenomenon are strategically crucial to the success of any business, including in the area of achieved benefits from mitigating the effects of adverse events occurring in the supply chain.

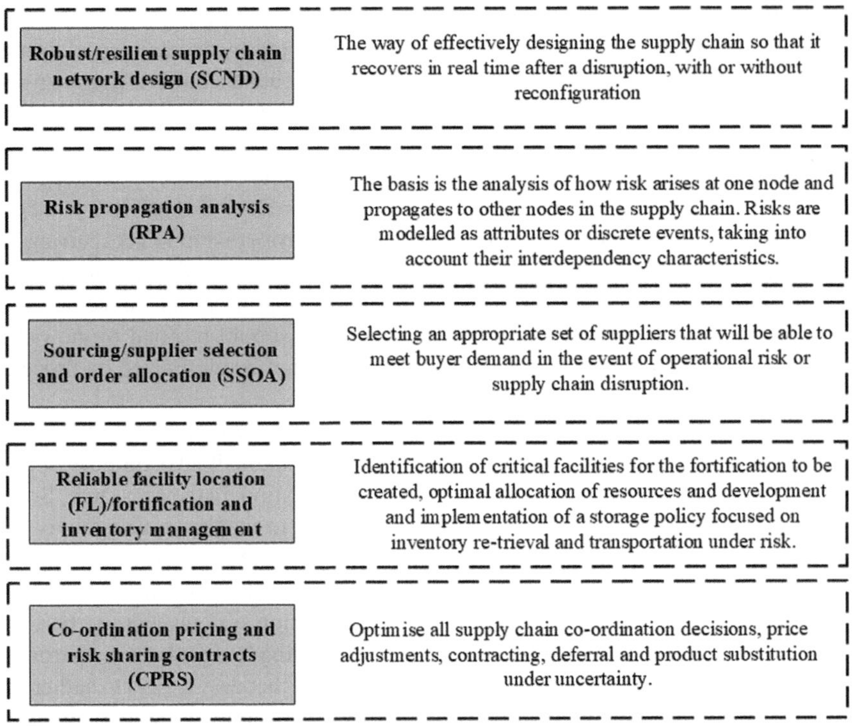

Fig. 2.8 Risk mitigation strategies (based on [50])

The growing interest in SCRM has resulted in several publications on systematic literature reviews in this area in recent years. Rajagopal et al. [50] distinguished the main research trends described in literature reviews on SCRM. These trends are shown in Fig. 2.9.

However, supply chain risk management is primarily a business issue, so the research described in [54], which presents a more practical approach, deserves attention. As a first step, the authors of this publication recommend categorising the risks present in the supply chain into two types [54]:

- known risks—these are risks that can be identified and measured based on quantitative and qualitative analysis. The probability and impact of these events can be estimated or quantified using information systems from the data provided for analysis. Interdisciplinary teams are set up to provide a risk management framework to identify the full range of these risks. This team can also identify areas where risks are difficult to understand and define;
- unknown risks—these are risks that cannot be predicted. Even the most experienced managers cannot predict the impact and possibility of occurrence scenarios.

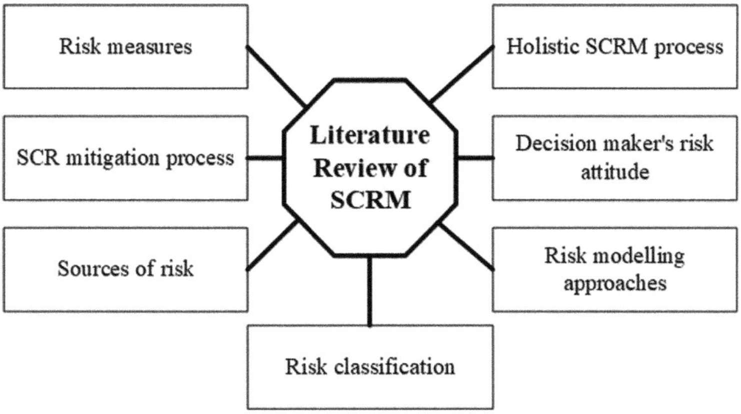

Fig. 2.9 Research trends analysed in literature reviews from the SCRM area (based on [50])

Increasing the speed of response to such risks is critical. For this reason, organisations are building layers of safeguards with a culture of risk awareness that can give them a competitive advantage.

Categorising risks into a specific group recommends different procedures for managing them. Figure 2.10 shows the pathways for known risks, and Fig. 2.11 for unknown risks.

2.1.2 Operational Risk Management

ERM is a strategic approach [18]. A similar point of view applies to most publications on the SCRM concept. However, from the point of view of the scope of the research, the risk management model at the operational level is relevant. Particular attention should be paid to the fact that, in the case of ERM and SCRM, the assessed risks can have a positive and a negative dimension. In the case of operational risk, the analysis carried out concerns only negative deviations [55].

The concept of operational risk is usually identified with the document produced by the Basel Committee of Banking Supervision, in which "operational risk is defined as the risk of loss resulting from inadequate or failed internal processes, people and systems or from external events. This definition includes legal risk but excludes strategic and reputational risk" [56]. The development of Basel II set out the requirements for operational risk management, which was seen as a new era in the development of the concept, and the definition of operational risk presented above has been accepted as a standard by financial organisations [57]. Based on this document, an operational risk framework was formulated based on the three pillars shown in Fig. 2.12.

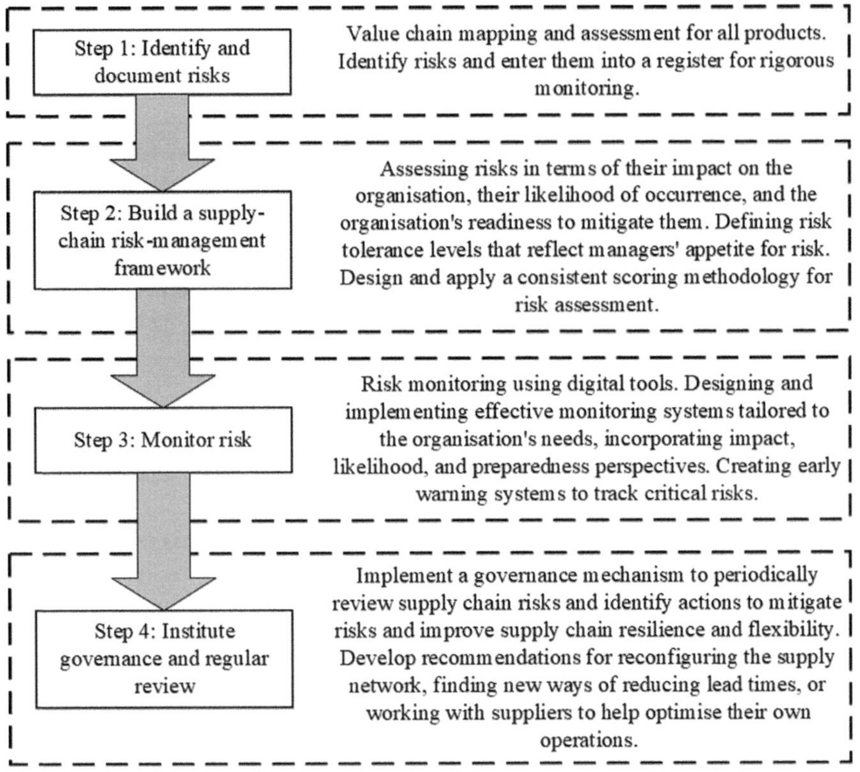

Fig. 2.10 Procedures for managing known risks (based on [54])

An operational risk policy is critical to effective risk management. It can form part of an overall risk management policy or stand-alone. Regardless of the form of drafting adopted, the policy should include [57]:

- The applicable definition of operational risk.
- The organisation's approach to operational risk management.
- The description of the main activities, including the roles and responsibilities of employees and managers.

It should also be noted that the implementation and promotion of an operational risk management culture should enable the achievement of the following organisational objectives as a minimum [55]:

- Members and internal stakeholders of the company understand the concept of operational risk.
- Employees know the procedures for dealing with operational risk.
- Managers recognise the benefits of good operational risk management.
- Managers are aware of the risks that occur without operational risk management.

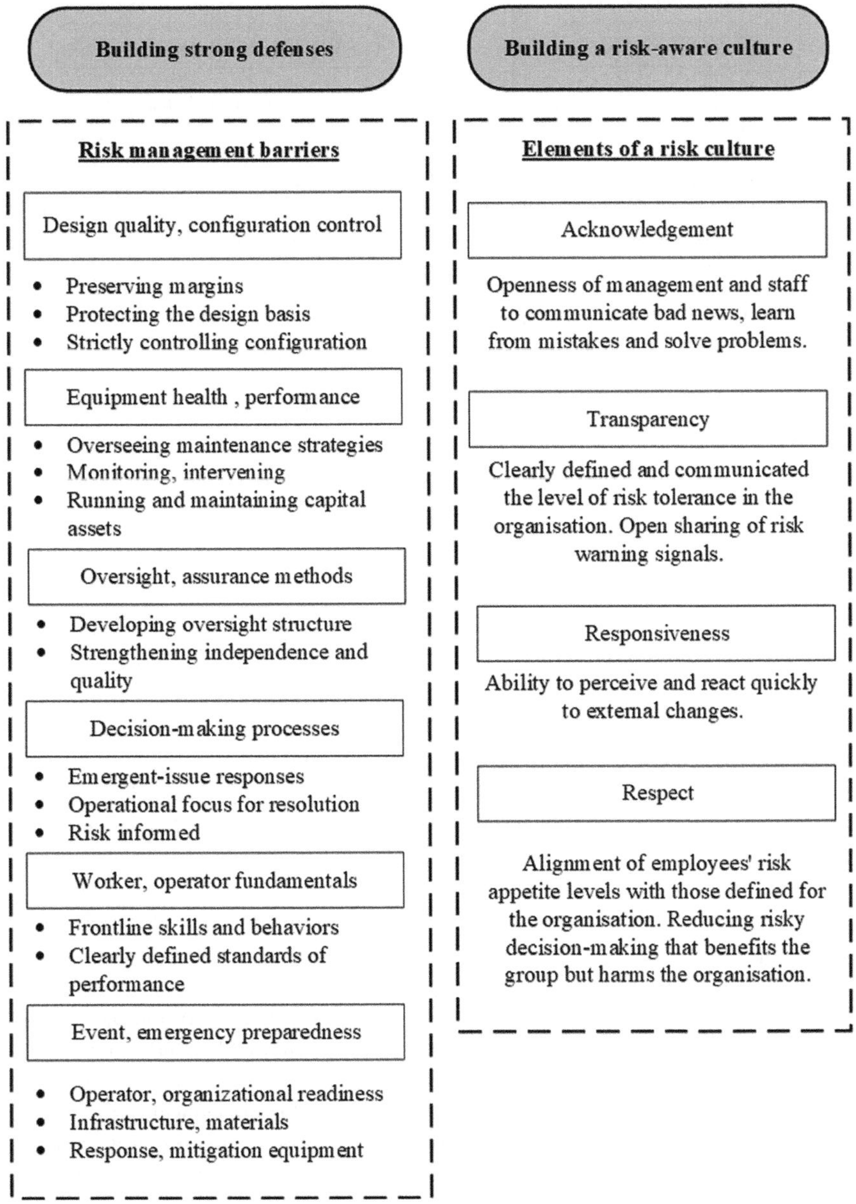

Fig. 2.11 Procedures for managing unknown risks (based on [54])

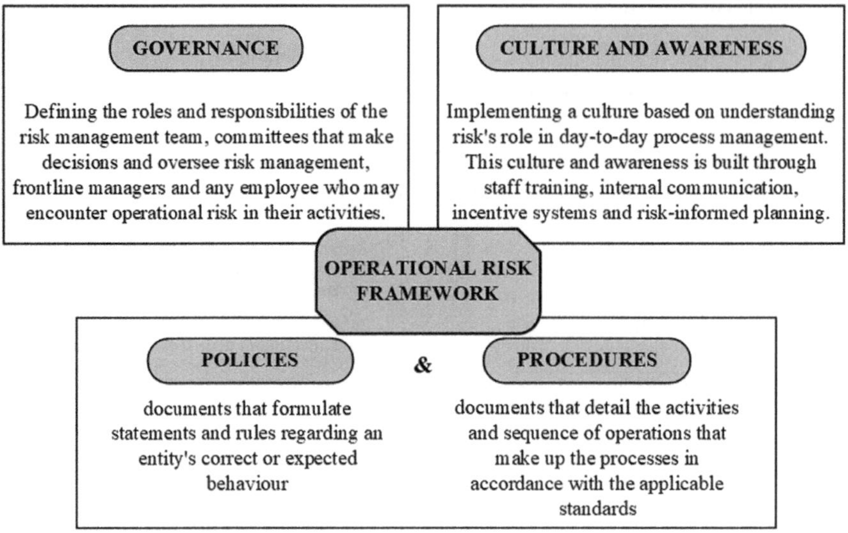

Fig. 2.12 Operational risk framework (based on [57])

- Key supporters of operational risk management have been identified, and a plan is in place to utilise their support.

In the document developed, the Basel Committee [56] has also distinguished four risk factors that can be present in any organisational unit and whose occurrence can cause losses of varying magnitudes. These factors are shown in Fig. 2.13.

Considering the risk factors highlighted by the Basel Committee as indicated above, seven categories of operational risk adverse events are usually distinguished in the literature. These categories are shown in Table 2.2.

In the financial services sector, operational risk is often referred to as non-financial risk, as opposed to credit, market or liquidity risk. This is because the risk factors can be people, systems or processes, so the sources of adverse events must be traced back to non-financial causes. In contrast, operational risk management primarily involves stabilising business outcomes, preventing incidents and accidents, and achieving the organisation's objectives in non-financial sectors.

Operational risk in non-financial organisations is also defined within the SCRM concept. When assessing operational risk in the supply chain, researchers focus primarily on disruptions that can negatively affect the performance of the entire supply chain. Among the causes of such disruptions, [59] is most often indicated:

- frequent changes in demand and supply along the chain;
- the impact of external factors or events on a supply chain partner, which then spreads to other partners;
- lack of transparency in the dependencies between chain partners.

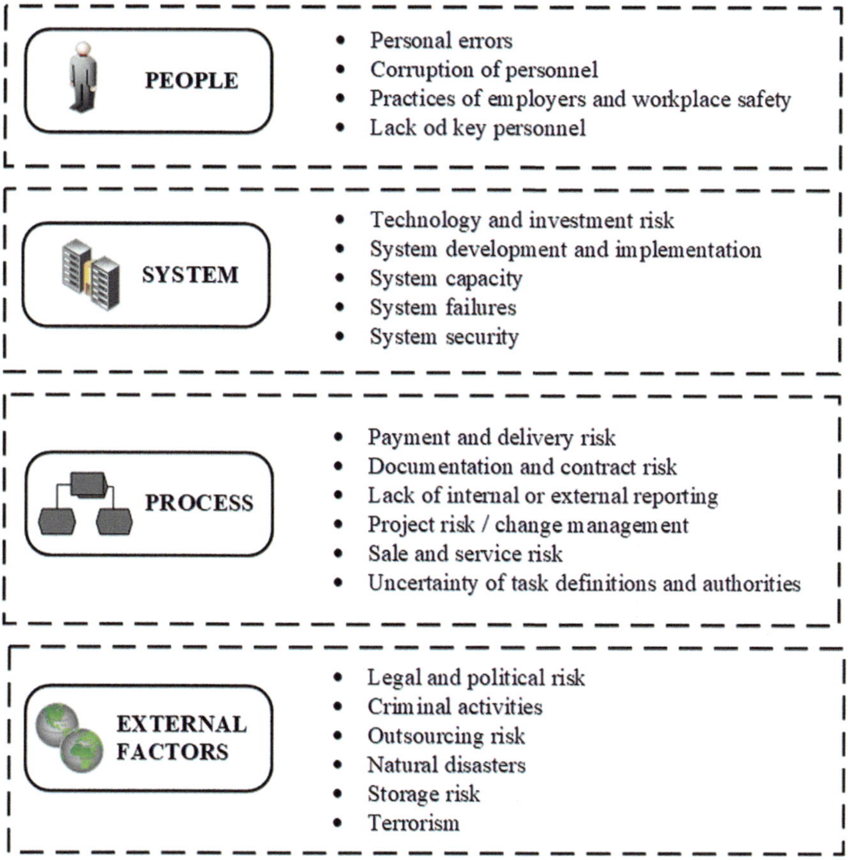

Fig. 2.13 Operational risk factors (based on [55])

A recommended approach to counteract the consequences of such disruptions is for supply chain partners to manage operational risks proactively. A proactive attitude in operational risk management means that each supply chain partner is focused on anticipatory detection of the occurrence of adverse events that may affect not only the operations of their organisation but also the functioning of the entire supply chain [59]. Partners detect disruptions occurring in internal processes and external phenomena, determine the probabilities of these disruptions and share this knowledge before developing plans to manage them.

Based on a literature review, Rajagopal et al. [50] proposed 11 categories of operational risk identified in the supply chain. These categories are characterised in Fig. 2.14.

The operational risk management model is similar to the models presented in ERM and SCRM. However, it is noteworthy that, according to [60], the starting point for the actions taken is the risk appetite. The critical issue in this case is translating

Table 2.2 Operational risk categories (based on [58])

Operational risk category	Examples of adverse events
Internal fraud	Losses result from inappropriate internal intentional or unintentional actions, such as non-compliance or violating laws, applicable procedures and regulations, and organisational policies. A minimum of one internal stakeholder must be involved
External fraud	Losses result from external intentional or unintentional actions, such as misuse of property, cybercrime, and infringing actions by third parties
Employment practices and workplace safety	Losses result from non-compliance with health and safety regulations or procedures in the processes carried out by the organization
Customers, products and business practices	Losses resulting from failure to meet obligations to customers, interference with mutual cooperation, as well as those caused by the nature or design of the product delivered to the market and business practices used
Damage to physical assets	Damage to physical assets caused by natural disasters, acts of terrorism or other events resulting in material loss
Business disruptions and system failures	Losses resulting from disruptions occurring in connection with business activities or system failures
Execution, delivery, process management	Strategies resulting from mismanagement of internal or external processes, including those related to disruptions in cooperation with business contractors

the appetite defined at the level of the organisation into the appetite of the business units. It should be remembered that risk appetite establishes the level of tolerance and acceptable risk limits, which influences the decision-making processes of business unit managers. Some authors even go a step further, stating that the desired risk appetite should be the basis of considerations for formulating organisational goals [61]. Therefore, the defined risk appetite initiates the operational risk management process shown in Fig. 2.15.

Risk managers in financial and non-financial organisations know that every disaster starts with an operational failure. Therefore, implementing a framework that provides transparency into the causes and consequences of potential adverse events at the operational level facilitates risk management decisions transparently and consistently across the enterprise. For this reason, several authors point to four core activities on which an effective operational risk management model framework is based. These activities are shown in Fig. 2.16.

These activities include [57]:

(1) Loss data collection

Operational risk management requires access to up-to-date data on events in and around the enterprise. Data from within the company refers to so-called internal

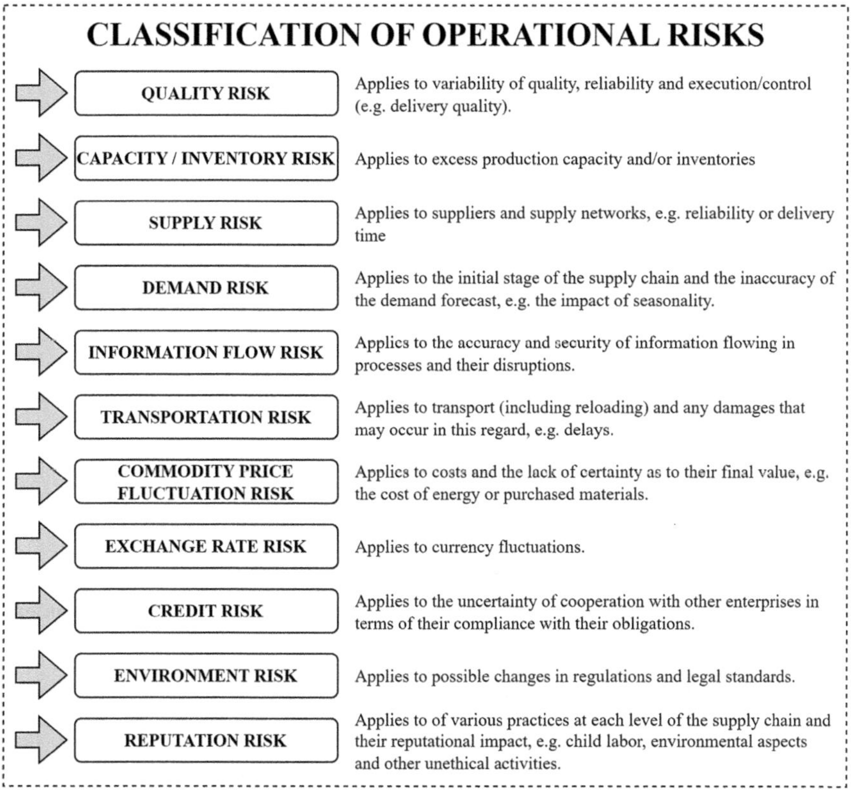

Fig. 2.14 Categories of operational risk in the supply chain (based on [50])

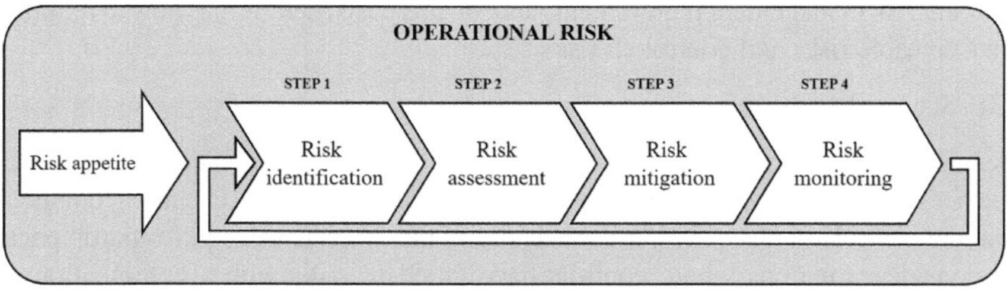

Fig. 2.15 Operational risk management model (based on [60])

losses, while industry data usually refers to external losses. A critical element of this activity is developing a compelling set of internal and external data. In the case of internal data collection, it is necessary to consider the organisational culture and risk awareness, which can affect the quality of the data provided. Operational risk events outside the organisation (in the industry or supply chain) are significant in understanding the risks the business faces, which it can influence minimally.

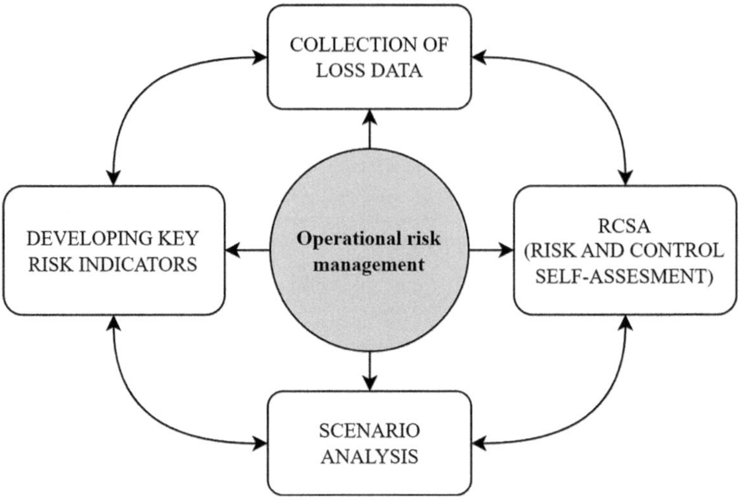

Fig. 2.16 Core operational risk management activities (based on [57])

Therefore, external data will be critical when creating scenario analyses and is integral to an effective reporting system.

(2) Risk and control self-assessment (RCSA)

Loss data refers to what has already happened, while RCSA is intended to help understand what potential risks may still occur. Its task is to assess all risks, not just those that have already occurred. The RCSA also aims to analyse current and future levels of risk, as its role is to predict future phenomena. It is considered by many authors to be the most critical part of the model, as it actively takes into account risk management requirements and, on this basis, determines how to mitigate unacceptable risks and control all risks.

(3) Scenario analysis

Scenario analysis focuses exclusively on rare events that have catastrophic consequences for the company. In RCSA, risk teams discuss the adverse events encountered and the controls in place. Scenario analysis, on the other hand, requires participants to consider situations where controls have failed or risks with a combination of parameters that have never been recorded before have occurred. Therefore, scenario analysis pushes team members to step outside the control zone and take a broad view of possible adverse events.

(4) Developing key risk indicators

Key risk indicators play a critical role in the risk monitoring process. They support the loss recording process and allow the individual RCSA stages to be actively supported. Their appropriate selection and the way they are recorded influence the effectiveness of the entire risk management process.

Developing a risk management model framework is critical, but its implementation within the company is equally important. For this reason, the research presented by Crouhy et al. [62] is noteworthy, as they highlighted eight key elements necessary to implement an operational risk management model successfully:

(1) *Developing well-defined operational risk policies*—defining the operational risk appetite; establishing minimum requirements for the internal control obligation that mitigate or reduce operational risk; identifying good practices identified from industry studies or in a group of comparable organisations; initiating empirical analyses to identify cause-and-effect relationships.
(2) *Establish standards for a common risk language*—a common risk language is necessary when conducting a qualitative or quantitative self-assessment by business management; it is essential for mutual understanding of the participants in different processes and for making operational risk decisions consistently and transparently across the enterprise.
(3) *Development of business process maps*—analysis of the organisation's products and services from the point of view of what is being done to manage operational risk; identification of causes of incidents and their detailed location in business processes; use of business process maps as input to the operational risk catalogue for all business in the organisation; categorisation and definition of the various operational risks including people, processes, systems and technology.
(4) *Develop a comprehensive set of operational risk indicators*—develop a quantitative methodology for measuring risk based on historical loss data and scenario analysis to identify the frequency distribution and severity of losses.
(5) *Define ways to manage operational risk exposure and mitigating actions*—analyse the cost–benefit ratio of taking targeted risk management actions, including risk transfer to another entity (insurance buy-outs); decisions should be supported by a portfolio that recommends solutions that maximise risk mitigation within available resources.
(6) *Identify ways to report risk exposures*—define operational risk numbers relevant to management and the board; create appropriate infrastructure to support reporting to relevant business units.
(7) *Develop analytical tools and procedures to determine when to apply these tools*—develop exposure measures; up-to-date databases of internal and industry operational loss data, scenario analyses, and ways to gather knowledge of critical causes of risk; establish risk frequency as a function of the degree to which a company's initiatives or evolving environment are expected to change risk.
(8) *The appropriate allocation of capital to each business.*

In conclusion, operational risk management must be based on a robust approach and a systematic process of integrating risk information and control into business decisions at the strategic and operational levels.

2.2 The Concept of Risk in the Engineering Approach

In the technical sciences, risk is seen only in negative terms, and its assessment refers to estimating the probability and consequences of a specific adverse event. The result of its occurrence can be loss or damage, the characteristics of which are shown in Fig. 2.17.

In the engineering approach, the concept of risk is linked to the orientation of safety management for the operation of technical systems. The development of the safety management concept was significantly influenced by the industrial revolutions, which changed the nature of work and increased the complexity of technical systems supporting human activities [64]. The first significant step in the evolution of the safety concept was the industry's attention to the need to provide adequate working conditions. The changes during the first decade of the Industrial Revolution highlighted the pressing need for a safe workspace for workers, the acceptance of which was ensured through the innovations of modern management pioneers. Industrial safety was also reinforced by the finding that it was more cost-effective and morally obligatory [65]. For this reason, many researchers in their publications emphasise that ensuring the safety of each person at home, at work, in transport, and throughout the surrounding environment is the shared responsibility of individual businesses, communities and government (e.g. [65]). Therefore, safety should be considered an aspect of the value system to be sought and nurtured [66].

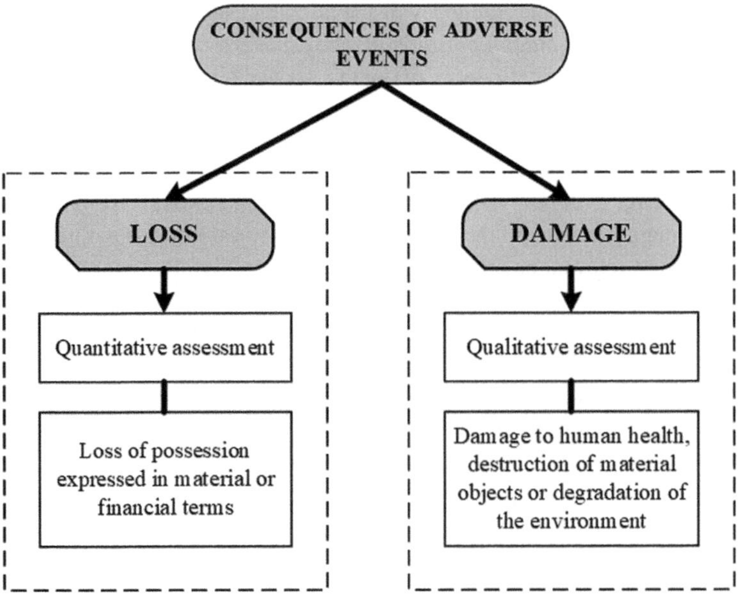

Fig. 2.17 Qualification of the consequences of an adverse event (based on [63])

Safety can already be considered positively and negatively, although the engineering approach views risk only negatively. Petterson and Deutsch characterised the two types of safety in their study as follows [67]:

(1) Safety—I. What goes wrong? This approach assumes that organisations conduct root cause analysis using various techniques. These analyses aim to identify factors that combine to lead to a specific adverse effect. Consequently, it is necessary to take action to identify the factors that contribute to an adverse outcome.
(2) Safety—II. What goes right? This approach is based on the principle that we should seek to understand and support more frequent occurrences of positive behaviour patterns. According to this approach, positive outcomes will occur as people adapt their behaviour to actual favourable working conditions.

In safety engineering, the first approach is predominant. This is why we can observe such a strong link between the concept of risk management in the engineering approach and research into the causes of hazardous situations. Hazard analysis is the heart of the developed safety systems [65]. However, it is worth pointing out that, in the case of anthropotechnical systems, the concept of hazard should not be restricted to the occurrence of failure (damage, error) alone. This is because a failure (error) describes a situation in which an object (human, machine, or other technical device) functions in an unintended manner (contrary to the intention of the decision-maker). At the same time, failure is not always accompanied by loss.

On the other hand, it is worth noting that undesirable events (accidents) may also occur during the correct operation of the object—i.e., without a failure. For this reason, it should be stressed that the concept of hazard is much broader in the context of risk assessment. It is defined as a potential condition or set of internal and external conditions for a system, product, facility or operator that, when activated, transforms the hazard into a series of events that end in loss [65]. Thus, it is essential to state that the term describes a state that can result in fatalities, injuries, destruction or loss of equipment/property, as well as harmful environmental effects.

Identifying and mitigating risks in the subsequent stages is a crucial cornerstone to support safety systems. Effective risk analysis throughout the life cycle of a technical system is the backbone to which all elements included in a safety programme are attached [65]. The concept of safety management thus calls for its analysis and monitoring of risk throughout the life cycle of a technical system, beginning with the conceptual phase of the system, continuing through the design, production, testing, operation, and finally, the disposal and discontinuation phases [65]. Different types of risks may be present in each phase of the system's life, and various characteristics and frequencies of their activation may be observed for the same risks, which also depend on the phase of the cycle. This is why it is crucial to continuously monitor technical systems for the risk of hazardous events and anticipate the risk-activating factors in a given phase of their life cycle.

Many authors emphasise that a safety system should be based on process and functional safety. According to the systematics presented in [68], process safety refers to all issues relating to the operation of technical systems in industrial sectors.

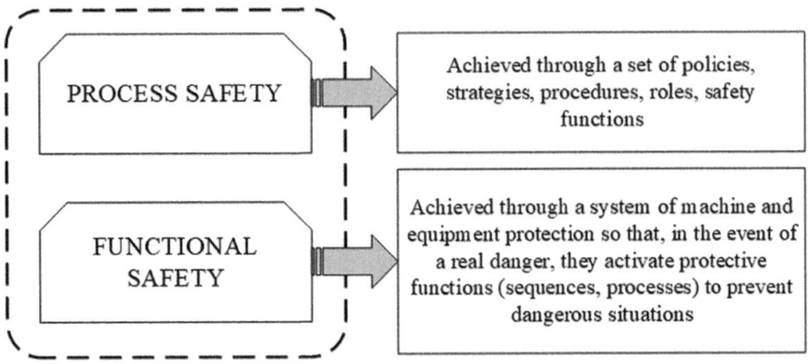

Fig. 2.18 Systematics of the relationship between process and functional safety (based on [68, 69])

On the other hand, functional safety is an engineering discipline that prevents hazards by appropriately designing safety features with specific functions. These systematics are illustrated in Fig. 2.18.

Process safety forms the basis for building a safety management system in an organisation. However, it is essential to remember that this system must be more than just a 'paper collection of policies and procedures' [70]. Safety management systems should be integrated mechanisms within an organisation designed to control hazards that affect the health and safety of workers [71]. For this reason, Fernandez-Muniz et al. in their research state that an adequate safety management system must include the following key areas [69]:

- The development of a safety policy, which includes a commitment by the organisation to provide an adequate level of safety and formally defines the objectives, principles, strategies, and guidelines to be followed in health and safety matters. It aims to create a climate that ensures that people from all levels of the organisation participate and contribute to building safety.
- Encouraging employees to engage in activities to enhance health and safety at work and promote safe behaviour.
- Creating a training and competence development system for employees to inform workers about risks and available prevention procedures. Its primary objective should be to change the attitudes of managers and workers so that they understand that safety is an essential part of doing the job properly and engage in continuous updating of their safety knowledge.
- Creating communication systems to efficiently communicate information about possible workplace hazards and the best ways to mitigate them.
- Distinguish between preventive and emergency planning. Preventive planning identifies possible hazards in the process environment, analyses the risk of their occurrence and proposes precautions to prevent accidents and losses. Contingency planning consists of preparing the human and material resources necessary to respond quickly and effectively in any emergency, limiting its potential effects as much as possible.

- Introducing a system of control and review of activities aimed at continuous improvement. The system distinguishes between two types of control—internal control and benchmarking analysis. Internal control will occur by analysing operations and working conditions (inspections and safety audits) and identifying and investigating incidents (accidents, incidents, near misses) occurring in the organisation. Benchmarking allows companies to compare their accident rates with other companies in the same sector and to evaluate the management techniques used (e.g. inspection procedures, accident investigation techniques, training programmes) against other organisations in any industry.

An analysis of the guidelines for the security management system reveals a high degree of correspondence between the highlighted elements and the guidelines for risk management systems in business terms. Therefore, for company processes supported by technical systems, integrating the safety management system with the risk management process is absolutely necessary.

2.3 The Interdisciplinary Approach to Operational Risk Assessment for Logistics Processes

The focus of the research presented in this book is operational risk concerning internal logistics processes (especially internal transport) supported by anthropotechnical systems. Using the approach to risk described in the concepts of ERM, SCRM and ORM, the Author has defined the concepts of operational risk to be used in the research described later in this book. The definition is presented in Table 2.3.

Internal logistics is responsible for handling material flows in the so-called internal supply chain. The scope of its service is shown in Fig. 2.19.

When assessing operational risks for internal logistics systems, it is crucial to consider the management perspective of the business processes that are the subject of the logistics service. The process approach compatibly involves managing material and information flows handled by logistics processes [73]. The main advantages of using BPM in the area of logistics include [74]:

- Real-time monitoring of the status of the business system.
- Ability to automate processes.
- Ability to change and improve the processes "on the fly".
- Ability to delegate rules to the process stakeholders.
- Benchmarking analysis of the processes.

Table 2.3 Definition of operational risk for the interdisciplinary approach (based on [72])

"Operational risk is the risk of an undesirable event happening, the occurrence of which disrupts the normal course of a process or reduces the achievement of an accepted operational objective."

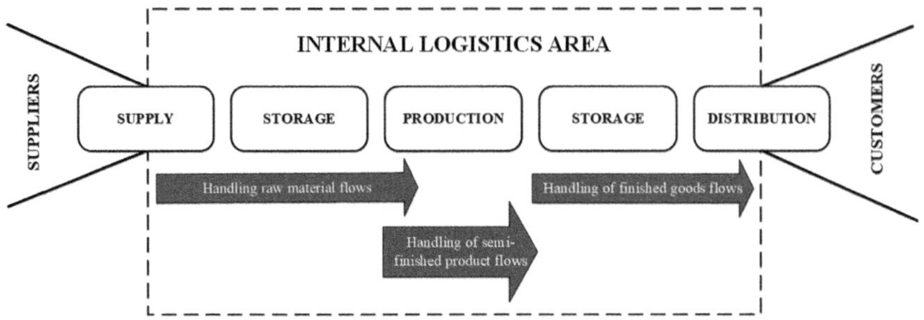

Fig. 2.19 Scope of service of internal logistics

The BPM approach also supports the risk management process itself. It identifies the risks associated with the activities undertaken and the resources required to carry them out. Adopting the BPM approach in assessing the risks associated with the operation of internal logistics systems allows a holistic perspective to be adopted for the analyses prepared. It combines elements of the business approach discussed in Sect. 2.1 with the engineering approach presented in Sect. 2.2.

The primary objective of internal logistics processes is to ensure the high efficiency and effectiveness of the company's goods flow handling processes. Therefore, although logistics processes are seen as so-called cost centres in most companies, risk management must not focus solely on cost aspects. These processes should be evaluated according to typical business process evaluation criteria based on time, cost and quality of results. Consequently, the risk management process should also refer to the triad defined in this way. The indicated parameters are shown in Fig. 2.20 concerning the evaluation of logistics processes in a company.

The indicated three criteria for improving internal logistics processes can be a starting point for risk management at the strategic level. However, above all, they are key measurement indicators for risk at the operational level. Those responsible for the risk assessment process will be referred to at the stage of analysing the potential impact of adverse events occurring, and the effectiveness of the risk mitigation measures implemented will be determined based on them. Also, these three criteria have a solid relationship, and a change in one of these parameters entails a simultaneous change in the others.

When carrying out a risk assessment for internal logistics processes, it should be noted that their implementation involves numerous trade-off relationships. These relationships should be an important part of the risk analysis but can also be used in mitigation. However, for this to be possible, it is necessary to balance the risks at the level of the entire logistics system at the assessment stage, as was done in the ERM concept at the company level. A high level of risk associated with stock availability may arise from the need to mitigate risks related to continuity of supply at another point in the process. This is why an integrated process approach to operational risk management for internal logistics is so important.

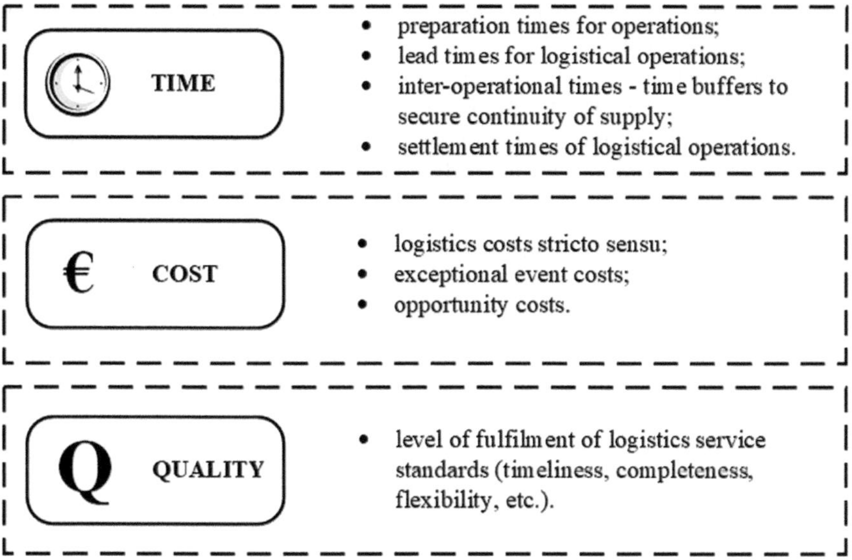

Fig. 2.20 Criteria for assessing internal logistics processes as a benchmark for operational risk management

An important feature of internal logistics processes is the overlap between movement and storage processes. Therefore, a critical aspect of their implementation is the proper coordination of material and information flows along the entire internal supply chain and their coordination and integration with other interdependent internal and external processes. This coordination should be carried out at the operational and managerial levels [75]:

- At the operational level, it means the coordination of processes, i.e. the agreement of mutual relations between activities or performers of activities, resulting in a dependence between them that is accepted by the performers, enabling the achievement of a specific defined objective or at least the improvement of the conditions for the performance of tasks resulting from this objective.
- At the managerial level, this means achieving a specific intended ordering of objectives, activities and processes, resulting from the adoption of an overarching business concept and specification of the desired results.

The final aim of coordination is to achieve unanimity and the required efficiency in implementing the activities that make up the process. To achieve this objective, it is necessary to define the relationships between the activities that make up the logistics processes. According to the classification presented in [76], it is essential to identify two basic types of relationships linking activities:

- Causal relations (causal)—the realisation of one activity is necessary for initiating the other; the result of one activity thus becomes the input resource of the other activity, e.g. the supply of material to the production station initiates the manufacturing process.

- Conditional relations—the realisation of one activity and the result obtained in it significantly influence the determination of the conditions for the commencement, realisation or termination of the other activity, e.g. the selection of a supplier determines the conditions for the realisation of the purchase of materials.

The lack of adequately identified core relationships results in the need for time and material buffers between activities and can generate disruptions in the implemented logistics flows. Therefore, their identification is critical from the point of view of operational risk management in internal logistics systems.

An important aspect of modern internal logistics processes is the need for them to be supported by technical systems. These systems may be of varying degrees of complexity and automation. However, their participation in the implemented operations generates the need to consider human–machine collaboration in the risk assessment (whereby the machine may be a device, robot or vehicle). Therefore, it is necessary to integrate an engineering approach into the risk management system for internal logistics and strongly orient the analyses carried out to the safety aspects of the operation of anthropotechnical systems. Therefore, the subject of the study will be incidents concerning damage to equipment supporting logistics operations and risks relating to abnormal human–machine interactions.

Taking into account the adopted BPM approach and the characteristics of the logistics processes, the Author developed a holistic approach to risk assessment for the internal logistics system in line with the framework presented in Fig. 1.1. Taking a holistic view of risk assessment at the stage of identifying adverse events requires a detailed analysis of the operations performed as part of the logistics service and the resources needed for their implementation. Identifying the process environment in which these operations are carried out is also essential. This environment can affect the results obtained. Therefore, the first risk assessment stage requires knowledge of the applicable procedures and internal regulations governing the manner (rules of execution) and sequence (relationships) of activities undertaken. The proposed process approach plays an important role here. One of the BPM procedure steps is mapping the current process flow and analysing its status, which forms the basis for further as-is assessment. At this stage of the process analysis, the critical role is played by managers and employees, whose knowledge and experience often determine the level of analysis carried out. Sources of information about the process will be the procedures in force (e.g. resulting from the implemented ISO standards), the scopes of responsibilities assigned to individual positions, document and information flow diagrams, and the comments, observations and habits of employee teams.

Managers should answer the questions presented in Fig. 2.21 when preparing process maps. An adequately prepared process map is the starting point for identifying interdisciplinary risks in internal logistics processes. It indicates the operations that make up the process and the resources involved in their implementation. This makes it possible to identify risks relating to the process and the systems supporting achieving the planned results.

- **Q1** What operations/activities are carried out in the process?
- **Q2** In what order are the individual actions performed?
- **Q3** Which operations cannot start before the others are completed?
- **Q4** What are the times of individual operations?
- **Q5** Are there any time buffers in the form of the so-called waiting time?
- **Q6** What people (departments) are involved in the implementation of individual activities?
- **Q7** What are the roles of the people involved in the implementation of each operation?
- **Q8** What resources are consumed during the implementation of the process?
- **Q9** What information is required by staff and managers to carry out the various operations?
- **Q10** What IT systems (digital solutions) support executing the selected operations?
- **Q11** What is the defined result of the process?

Fig. 2.21 Basic questions posed to managers at the stage of analysing the current state of the process

The questions formulated in Fig. 2.21 indicate that an important issue at the risk identification stage in anthropotechnical systems is knowledge of the technical solutions that support the execution of individual operations. From the point of view of risk assessment, information on their functionality, availability, and performance is essential, as operational objectives for the technical system supporting the logistics processes will be formulated based on these parameters. From the point of view of safety and continuity of operation, it is also important to know the technical equipment's accepted service standards and operating conditions. Modern logistics processes are increasingly supported by complex cyber-physical systems whose operational reliability guarantees the continuity of process execution. For this reason, identifying risks associated with disrupting the machinery, equipment, and vehicles in use is essential to the analysis prepared for logistics processes.

Among the resources required for implementing activities in anthropo-technical systems, teams of employees and their managers play a crucial role. Therefore, for risk assessment purposes, knowing what human resources are available and considering their number, qualifications, and experience is necessary. Deficiencies relating to these three parameters will generate disruptions in the logistical processes and may affect the occurrence of deviations from the formulated target results. The working conditions and their current psycho-physical state must also be taken into account when identifying adverse events involving people. Many errors made by people involved in processes are due not only to a lack of knowledge or experience but also to stress, fatigue or poor health. Managers should carefully analyse all these elements when identifying adverse events. This is because human errors can lead to incorrect results of the performed action but also cause damage to technical systems.

Available information is also an important resource affecting the occurrence of adverse events. Their absence or inadequate quality is a critical source of possible disruptions. Therefore, at the risk identification stage, it is necessary to know the available databases and how they are collected, processed and distributed to implement logistics operations. It is also worth characterising the IT systems supporting the planning of logistical processes and the implementation of selected operations. Concerning logistics processes, in addition to integrated planning systems (MRP, ERP, WMS), companies are keen to use systems that support quantitative analyses, e.g. simulation programmes, business intelligence systems, and automatic data identification systems. These systems are so widely used that even a temporary reduction in their availability causes critical disruptions in the implementation of logistics processes. Therefore, at the risk identification stage, verification of handling systems and safeguards that protect collected data and guarantee continuity of information handling of processes becomes an important aspect.

When preparing the list of possible events, it is essential to remember to include both historical data from the company's internal information systems, phenomena occurring in similar supply chains and events not yet recorded but whose probability of occurrence increases due to changing internal and external conditions affecting process execution. The list of potential risks should include all events that, if they occur, will result in the failure or limitation of achieving the defined objectives for a given logistics process.

The second stage of work focuses on risk analysis. The aim of this stage is primarily to estimate measures of risk—the probability and consequences of an adverse event—and to determine the value of a risk indicator. The study of cause-and-effect relationships may also complement this stage, the detailed analysis of which allows for a better understanding of the phenomena characterising a given adverse event and, thus, a better assessment of the potential consequences of its occurrence. The ISO 31000:2018 standard recommends quantitative and qualitative methods used at different risk assessment stages. A critical aspect of the analyses performed is the availability of quantitative and qualitative data that meet the requirements defined for the inputs of the analytical model. Quantitative or quantitative–qualitative methods are recommended to estimate measures of known risk. These methods will determine the level of risk in quantitative terms based on historical data. Scenario analysis is recommended to assess the level of unknown risk. Its results can be expressed in terms of a risk index and a linguistic variable describing the level of risk.

The final stage of risk assessment is risk evaluation. A critical element of this stage is determining risk acceptance levels based on the managers' risk appetite as defined for a given logistics system. Distinguishing between different risk acceptance levels allows diversification of the actions taken to mitigate risk and the time–space of their implementation. Based on the formulated risk acceptance levels, a risk matrix is created to prioritise implementing mitigation actions and prepare proactive and reactive risk handling plans.

The results of analyses of the causes of adverse events are also used in the risk evaluation stage. This is because their identification allows an organisation's degree of control over a given risk to be determined. Such a classification is essential from the point of view of the risk management process because [72]:

- the level of control exercised determines the choice of risk control mechanisms;
- the extent of control exercised over a given element defines the potential for process improvement by the enterprise within the concept of continuous improvement;
- the level of control exercised defines the possibility of influencing the factors that activate a given risk.

Internal resources and operating procedures are mainly supervised and controlled by the enterprise. As a result, they can be developed and improved in line with the organisation's change implementation plan, and the risks involved can thus be effectively mitigated. From the perspective of logistics process management, these events also take on particular importance. By identifying and classifying them, it is possible to introduce continuous improvement programmes aimed at:

- more efficient use of company resources;
- elimination of existing wastage;
- increasing the level of logistics services provided while optimising the associated costs;
- eliminating internal disruptions that affect the process results.

Concerning risks that originate from external factors, the level of control exercised by the organisation, and therefore the possibility of influencing the risk, is limited. For this reason, at the operational risk management level, managers can implement reporting systems that allow them to continuously monitor environmental situations to consider changes in the prepared action plans [72].

The risk evaluation results should form the basis for operational planning of the logistical handling of internal material flows but also provide input for the preparation of change programmes and improvements to the organisation's internal processes.

2.4 Summary

The concept of risk is understood differently, and many approaches to risk assessment and management can be found in the literature. The aim of this chapter was to present the most popular risk management concepts described in the literature and applied in practice over the last 20 years, which also formed the basis for the developed risk management approach for internal transport. However, the scope of the concepts described has been limited to those solutions used in industry, which aligns with the research perspective adopted in this book.

In Sect. 2.1, three leading research trends on business approaches to risk management are characterised. The most crucial concept in this case is the Enterprise Risk Management strategy, which has transformed how risk is approached in companies and its use in decision-making processes. Thanks to the standards developed by the COSO organisation, the notion of risk culture and risk portfolio management has been reinforced in business terminology, and the processes involved in assessing and dealing with risks have been given due prominence. The ERM concept has allowed organisations to move away from the siloed approach to risk assessment and management and highlighted the need to integrate it at the enterprise level. Also critical is the dual approach to risk promoted by the ERM concept, in which the impact of uncertainty on objectives can be assessed both positively and negatively. In the traditional approach to risk management, managers' attention was mainly focused on protecting the company through the adverse financial consequences associated with the occurrence of risk. In the ERM concept, risk management is an important strategic element, and using the results of the risk analyses is expected to influence more effective and efficient decision-making processes in the organisation.

Based on the ERM concept, the ISO 31000 standard was developed and implemented in industry, which recommends a framework for managing risk in organisations based on four basic processes including (a) risk identification, (b) risk analysis, (c) risk evaluation, (d) risk treatment; and two supporting processes: (a) monitoring and review and (b) communication and consultation. For the process structure defined in this way, principles have also been developed to effectively and efficiently manage risks to improve organisational performance. The standards proposed in the ISO 31000:2018 framework are now being implemented in different types of organisations and all sectors of the economy. The developed framework also provides a basis for further developing the concept of risk management in companies. The framework recommended by ISO 31000:2018 also forms the basis for the concept described in this book.

A second significant research trend that needs to be considered is the concept of Supply Chain Risk Management. Its assumptions are based on the approach presented in the ERM strategy and the ISO 31000 standard. However, its analyses consider the level of individual organisations and the risks associated with the functioning of the entire supply chain. For this reason, risk assessment targets micro or macro events whose occurrences affect individual links or the entire supply chain. At the same time, most researchers in this area stress that, when identifying adverse events in the supply chain, they should be classified into internal, external, and network risk. This classification will determine the level of control exercised over the identified adverse events and, consequently, identify potentially feasible actions to mitigate the risk.

A critical aspect of the SCRM concept is adopting an individual enterprise (SC link) and an entire supply chain perspective in risk assessment. Indeed, the success of its implementation is determined by managers' understanding that the risk vulnerability of an enterprise is not only a question of risk management at the level of the individual organisation but is also highly dependent on the risks of other supply chain participants. Therefore, it is necessary to strengthen the logistics systems handling internal and external flows in the individual links of the supply chain, as this is the only way to achieve the required level of resilience for the entire supply chain. The supply chain will only be as vulnerable to disruptions as its weakest link. In addition, it should be stressed that, in the case of supply chain risks, a domino effect and a snowball effect can be observed. This means that a minor disruption occurring at one actor in the supply chain can generate critical adverse events for the entire chain. Hence, the need arises to identify adverse events at the strategic and operational levels, as only in this way is its timely mitigation possible.

For this reason, the third area of research on risk in business terms was the concept of Operational Risk Management. This concept refers in most publications to a document developed by the Basel Committee of Banking Supervision, in which risk is defined as a loss caused by a person, system, process or external factors. What should be emphasised is that in this approach, risk already has an exclusively negative dimension and is identified with events whose occurrence causes loss to the company. Operational risk, described by the Basel Committee of Banking Supervision, refers exclusively to financial institutions. However, within a short time, this approach was adapted to the specifics of non-financial organisations. Also, within the SCRM concept, some authors identify the need for risk management at the operational level and, for identified adverse events at this level, recommend proactive attitudes aimed at detecting in advance the possibility of their occurrence by all partners in the supply chain.

According to the perspective of logistics systems risk assessment, it is crucial to complement the business view with an engineering approach to risk management. For this reason, Sect. 2.2 characterises risks relating to technical systems, which in engineering science are only assessed in terms of adverse outcomes expressed in terms of loss (quantitative view) or damage (qualitative view). In the engineering approach, risk assessment is linked to safety management and analysing hazards in anthropotechnical systems. The hazard is described as a potential system condition or set of conditions that may result in fatalities, injuries, destruction or loss of equipment/property, and harmful environmental effects. At the same time, research indicates that different risks may be present in each phase of the life cycle of a technical system, and the same hazards may have different characteristics and activation frequencies, which depend on the phase of the cycle.

Based on the presented business and engineering approaches to risk management, an interdisciplinary approach to operational risk assessment for internal logistics processes was developed. Risk assessment is concerned with the operational level, as the analysis of risks is directed towards achieving the goal of providing the expected level of logistics service. The efficiency and effectiveness of achieving this goal depend mainly on the correct coordination of activities to ensure smooth (uninterrupted) execution of individual operations. To achieve this, it is therefore necessary to adopt a process perspective. For this reason, a concept that supports the operational risk assessment process in this case is the BPM approach, which makes it possible to identify the risks associated with the activities undertaken and the resources used to carry them out. The proposed approach also takes into account the fact that anthropotechnical systems support modern internal logistics processes. For this reason, aspects concerning assessing the operational safety of technical systems that support the realisation of logistics processes are included in the analyses conducted. On this basis, four evaluation criteria are recommended for investigating adverse events' consequences. These are time, cost, quality and impact on the health and lives of process participants.

The concept of risk assessment for internal logistics processes presented in Sect. 2.3 uses some of the assumptions formulated for the strategy framework described in Sects. 2.1 and 2.2. It is also a response to the needs of contemporary logistics processes, the effectiveness and efficiency of which are often a source of competitive advantage for both individual companies and entire supply chains. The concept also fills a current research gap: the lack of operational risk assessment models dedicated to the needs of logistics processes while considering a managerial and engineering approach to process analysis. Above all, however, the framework for risk assessment described in this chapter is the basis for a new approach to risk management relating to the internal transport process supported by Logistics 4.0 systems.

References

1. Bromiley P, McShane M, Nair A, Rustambekov E (2015) Enterprise risk management: review, critique, and research directions. Long Range Plann 48:265–276. https://doi.org/10.1016/j.lrp.2014.07.005
2. Tubis AA, Poturaj H (2022) Risk related to AGV systems—open-access literature review. Energies (Basel) 15:8910. https://doi.org/10.3390/en15238910
3. Tubis AA, Poturaj H (2021) Challenges in the implementation of autonomous robots in the process of feeding materials on the production line as part of Logistics 4.0. Logforum 17:411–423. https://doi.org/10.17270/J.LOG.2021.611
4. Tubis AA, Rohman J (2023) Intelligent warehouse in Industry 4.0—systematic literature review. Sensors 23:4105. https://doi.org/10.3390/s23084105
5. Tubis AA, Ryczyński J, Żurek A (2021) Risk assessment for the use of drones in warehouse operations in the first phase of introducing the service to the market. Sensors 21:6713. https://doi.org/10.3390/s21206713
6. Tubis AA, Poturaj H, Smok A (2024) Interaction between a Human and an AGV system in a shared workspace—a literature review identifying research areas. Sustainability 16:974. https://doi.org/10.3390/su16030974
7. Stulz RM (1996) Rethinking risk management. J Appl Corp Financ 9:8–25. https://doi.org/10.1111/j.1745-6622.1996.tb00295.x

8. Liebenberg AP, Hoyt RE (2003) The determinants of enterprise risk management: evidence from the appointment of chief risk officers. Risk Manage Insurance Rev 6:37–52. https://doi.org/10.1111/1098-1616.00019
9. Sobel PJ, Reding KF (2004) Aligning corporate governance with enterprise risk management. Manag Account Q 5:29–37
10. Committee of sponsoring organizations of the treadway commission: enterprise risk management—integrated framework
11. Committee of sponsoring organization of the treadway commission: enterprise risk management—integrating strategy with performance
12. Rubino M (2018) A comparison of the main ERM frameworks: how limitations and weaknesses can be overcome implementing IT governance. Int J Bus Manage 13:203. https://doi.org/10.5539/ijbm.v13n12p203
13. Dickinson G (2001) Enterprise risk management: its origins and conceptual foundation. Geneva Pap Risk Insur Issues Pract 26:360–366. https://doi.org/10.1111/1468-0440.00121
14. Harrington SE, Niehaus G, Risko KJ (2002) Enterprise risk management: the case of united grain growers. J Appl Corp Financ 14:71–81. https://doi.org/10.1111/j.1745-6622.2002.tb00450.x
15. Meulbroek LK (2002) Integrated risk management for the firm: a senior manager's guide. SSRN Electronic J. https://doi.org/10.2139/ssrn.301331
16. Arena M, Arnaboldi M, Azzone G (2010) The organizational dynamics of enterprise risk management. Acc Organ Soc 35:659–675. https://doi.org/10.1016/J.AOS.2010.07.003
17. Blanco-Mesa F, Rivera-Rubiano J, Patiño-Hernandez X, Martinez-Montaña M (2019) The importance of enterprise risk management in large companies in Colombia. Technol Econ Dev Econ:1–34. https://doi.org/10.3846/tede.2019.9380
18. Soltanizadeh S, Rasid SZA, Golshan N, Quoquab F, Basiruddin R (2014) Enterprise risk management practices among Malaysian firms. Procedia Soc Behav Sci 164:332–337. https://doi.org/10.1016/j.sbspro.2014.11.084
19. Institute of Risk Management: From the cube to the rainbow double helix: a risk practitioner's guide to the COSO ERM Frameworks, London (2018)
20. Renault BY, Agumba JN, Balogun OA (2016) Drivers for and obstacles to enterprise risk management in construction firms: a literature review. Procedia Eng 164:402–408. https://doi.org/10.1016/j.proeng.2016.11.637
21. Alijoyo FA, Hendra R, Sirait KB (2021) The State-of-the-Art of enterprise risk management maturity models: a review. Ann Romanian Soc Cell Biol 25:4005–4014
22. Anton SG, Nucu AEA (2020) Enterprise risk management: a literature review and agenda for future research. J Risk Fin Manage 13:281. https://doi.org/10.3390/jrfm13110281
23. ISO 31000:2018 Risk management—guidelines
24. Alijoyo FA, Norimarna S (2021) Risk management maturity assessment based on ISO 31000—a pathway toward the organization's resilience and sustainability post COVID-19: the case study of SOE company in Indonesia. In: Proceedings of the 3rd international conference on management, economics and finance. ACAVENT, pp 125–142. https://doi.org/10.33422/3rd.icmef.2021.02.134
25. Oliva FL (2016) A maturity model for enterprise risk management. Int J Prod Econ 173:66–79. https://doi.org/10.1016/j.ijpe.2015.12.007
26. Tubis AA, Werbińska-Wojciechowska S (2021) Risk management maturity model for logistic processes. Sustainability 13:659. https://doi.org/10.3390/su13020659
27. Rubino M, Vitolla F (2014) Internal control over financial reporting: opportunities using the COBIT framework. Manag Audit J 29:736–771. https://doi.org/10.1108/MAJ-03-2014-1016
28. Chopra S, Sodhi MS (2004) Managing risk to avoid supply chain breakdown. MIT Sloan Manag Rev 46:53–62
29. Hallikas J, Karvonen I, Pulkkinen U, Virolainen VM, Tuominen M (2004) Risk management processes in supplier networks. Int J Prod Econ 90:47–58. https://doi.org/10.1016/J.IJPE.2004.02.007

30. Zsidisin GA (2003) A grounded definition of supply risk. J Purch Supply Manag 9:217–224. https://doi.org/10.1016/J.PURSUP.2003.07.002
31. Peck H (2006) Reconciling supply chain vulnerability, risk and supply chain management. Int J Log Res Appl 9:127–142. https://doi.org/10.1080/13675560600673578
32. Ho W, Zheng T, Yildiz H, Talluri S (2015) Supply chain risk management: a literature review. Int J Prod Res 53:5031–5069. https://doi.org/10.1080/00207543.2015.1030467
33. Wieland A, Wallenburg CM (2012) Dealing with supply chain risks. Int J Phys Distrib Logist Manag 42:887–905. https://doi.org/10.1108/09600031211281411
34. Manuj I, Mentzer JT (2008) Global supply chain risk management strategies. Int J Phys Distrib Logist Manag 38:192–223. https://doi.org/10.1108/09600030810866986
35. Tang CS (2006) Perspectives in supply chain risk management. Int J Prod Econ 103:451–488. https://doi.org/10.1016/j.ijpe.2005.12.006
36. Fan Y, Stevenson M (2018) A review of supply chain risk management: definition, theory, and research agenda. Int J Phys Distrib Logist Manag 48:205–230. https://doi.org/10.1108/IJPDLM-01-2017-0043
37. Blos MF, Wee HM, Yang W-H (2012) Supply chain risk management: resilience and business continuity. In: Lu J, Jain LC, Zhang G (eds) Handbook on decision making. Intelligent systems reference library. Springer, Berlin, pp 219–236. https://doi.org/10.1007/978-3-642-25755-1_12
38. Christopher M, Peck H (2004) Building the resilient supply chain. Int J Logis Manage 15:1–14. https://doi.org/10.1108/09574090410700275
39. Jallow AK, Majeed B, Vergidis K, Tiwari A, Roy R (2007) Operational risk analysis in business processes. BT Technol J 25:168–177. https://doi.org/10.1007/s10550-007-0018-4
40. Kendrick T (2009) Identifying and managing project risk: essential tools for failure-proofing your project. Amacom Books
41. Venkatesh VG, Rathi S, Patwa S (2015) Analysis on supply chain risks in Indian apparel retail chains and proposal of risk prioritization model using interpretive structural modeling. J Retail Consum Serv 26:153–167. https://doi.org/10.1016/j.jretconser.2015.06.001
42. Sarker S, Engwall M, Trucco P, Feldmann A (2016) Internal visibility of external supplier risks and the dynamics of risk management silos. IEEE Trans Eng Manag 63:451–461. https://doi.org/10.1109/TEM.2016.2596144
43. Wagner SM, Bode C (2008) An empirical examination of supply chain performance along several dimensions of risk. J Bus Logist 29:307–325. https://doi.org/10.1002/j.2158-1592.2008.tb00081.x
44. Mishra D, Sharma RRK, Kumar S, Dubey R (2016) Bridging and buffering: Strategies for mitigating supply risk and improving supply chain performance. Int J Prod Econ 180:183–197. https://doi.org/10.1016/J.IJPE.2016.08.005
45. Talluri S (Sri), Kull TJ, Yildiz H, Yoon J (2013) Assessing the efficiency of risk mitigation strategies in supply chains. J Bus Logis 34:253–269. https://doi.org/10.1111/jbl.12025
46. Bode C, Wagner SM, Petersen KJ, Ellram LM (2011) Understanding responses to supply chain disruptions: insights from information processing and resource dependence perspectives. Acad Manag J 54:833–856. https://doi.org/10.5465/amj.2011.64870145
47. Chattopadhyay P, Glick WH, Huber GP (2001) Organizational actions in response to threats and opportunities. Acad Manag J 44:937–955. https://doi.org/10.2307/3069439
48. Raj Sinha P, Whitman LE, Malzahn D (2004) Methodology to mitigate supplier risk in an aerospace supply chain. Supply Chain Manage: Int J 9:154–168. https://doi.org/10.1108/13598540410527051
49. Premkumar G, Ramamurthy K, Saunders CS (2005) Information processing view of organizations: an exploratory examination of fit in the context of interorganizational relationships. J Manag Inf Syst 22:257–294. https://doi.org/10.1080/07421222.2003.11045841
50. Rajagopal V, Prasanna Venkatesan S, Goh M (2017) Decision-making models for supply chain risk mitigation: a review. Comput Ind Eng 113:646–682. https://doi.org/10.1016/j.cie.2017.09.043

51. Wang L, Yeung JHY, Zhang M (2011) The impact of trust and contract on innovation performance: the moderating role of environmental uncertainty. Int J Prod Econ 134:114–122. https://doi.org/10.1016/j.ijpe.2011.06.006
52. Yeung JHY, Selen W, Zhang M, Huo B (2009) The effects of trust and coercive power on supplier integration. Int J Prod Econ 120:66–78. https://doi.org/10.1016/j.ijpe.2008.07.014
53. Handley SM, Benton WC (2012) Mediated power and outsourcing relationships. J Oper Manag 30:253–267. https://doi.org/10.1016/j.jom.2011.11.004
54. Bailey T, Barriball E, Dey A, Sankur A. A practical approach to supply-chain risk management
55. Büyüközkan G, Göçer F (2018) Digital supply chain: literature review and a proposed framework for future research. Comput Ind 97:157–177. https://doi.org/10.1016/j.compind.2018.02.010
56. Basel Committee on Banking Supervision (BCBS) Basel II: International convergence of capital measurement and capital standards: a revised framework—comprehensive version
57. Girling P (2022) Operational risk management: a complete guide for banking and Fintech. John Wiley & Sons Inc., Hoboken, New Jersey
58. Pieket Weeserik B, Spruit M (2018) Improving operational risk management using business performance management technologies. Sustainability 10:640. https://doi.org/10.3390/su10030640
59. Nimmy SF, Hussain OK, Chakrabortty RK, Hussain FK, Saberi M (2022) Explainability in supply chain operational risk management: a systematic literature review. Knowl Based Syst 235:107587. https://doi.org/10.1016/j.knosys.2021.107587
60. PWC operational risk management
61. Samad-Khan A (2008) Enterprise risk management modern operational risk management. Emphasis 2:26–29
62. Crouhy M, Galai D, Mark R (2023) The essentials of risk management. McGraw Hill
63. Smalko Z, Nowakowski T, Tubis A (2020) Framework for reliability hazard theory (in Polish). Wroclaw University of Science and Technology Publishing House, Wrocław
64. Hendrick HW (1991) Ergonomics in organizational design and management. Ergonomics 34:743–756. https://doi.org/10.1080/00140139108967348
65. Harold E (1991) Roland, Brian Moriarty: system safety engineering and management. John Wiley & Sons Inc., New York/Chichester/Toronto
66. Czupryński A, Wiśniewski B, Zboina J (2015) Safety. Theory—research—practice (in Polish). CNBOP-PIB. https://doi.org/10.17381/2015.4
67. Patterson M, Deutsch ES (2015) Safety-I, safety-II and resilience engineering. Curr Probl Pediatr Adolesc Health Care 45:382–389. https://doi.org/10.1016/j.cppeds.2015.10.001
68. Kozak A (2011) Process safety in industrial facilities (in Polish). Budownictwo i Inżynieria Środowiska. 2:319–322
69. Fernández-Muñiz B, Montes-Peón JM, Vázquez-Ordás CJ (2007) Safety management system: Development and validation of a multidimensional scale. J Loss Prev Process Ind 20:52–68. https://doi.org/10.1016/j.jlp.2006.10.002
70. Mearns K, Whitaker SM, Flin R (2003) Safety climate, safety management practice and safety performance in offshore environments. Saf Sci 41:641–680. https://doi.org/10.1016/S0925-7535(02)00011-5
71. Labodová A (2004) Implementing integrated management systems using a risk analysis based approach. J Clean Prod:571–580. https://doi.org/10.1016/j.jclepro.2003.08.008
72. Tubis A (2018) Operational risk management method in road transport (in Polish). Wroclaw University of Science and Technology Publishing House, Wrocław
73. Božić D, Stanković R, Rogić K (2014) Possibility of applying business process management methodology in logistic processes optimization. PROMET—Traffic Transp 26:507–516. https://doi.org/10.7307/ptt.v26i6.1610
74. Wiechetek Ł, Mędrek M (2019) Synergizing the logistics processes and process management—framework of BPM course based on business cases, BPM standards and tools. Annales Universitatis Mariae Curie-Skłodowska, sectio H—Oeconomia 52:95. https://doi.org/10.17951/h.2018.52.5.95-112

75. Krawczyk S (2011) Logistics. Theory and practice (in Polish). Difin, Warsaw
76. Krawczy S (2016) Concepts of the universal in research (in Polish). University of Zielona Gora Publishing House, Zielona Gora

Chapter 3
The Impact of Industry 4.0 on Logistics Systems

Contents

Chapter 1 noted that the digital transformation observed for many years has significantly impacted the growing interest in the concept of risk management, increasing the possibilities regarding the collection and analysis of adverse event data and its use in adopting proactive attitudes towards identified risks. The increasingly rapid digital transformation has also supported the development of the so-called fourth industrial revolution (Industry 4.0). Industry 4.0, like all industrial revolutions, has introduced significant changes in manufacturing processes, technologies and systems.

A. Tubis, *The New Paradigm of Risk in Internal Transport Supporting Logistics 4.0 System*, Lecture Notes in Intelligent Transportation and Infrastructure,
https://doi.org/10.1007/978-3-031-64920-2_3

However, its influence very quickly began to affect other processes around production, including internal logistics processes. Solutions, initially dedicated primarily to manufacturing systems, very quickly began to develop to support logistics processes as well. Therefore, with the development of Industry 4.0, the concept of Logistics 4.0 emerged, which uses cyber-physical solutions to support material flows. Today, Logistics 4.0 is considered one of the critical trends of Industry 4.0.

3.1 Development of the Industry 4.0 (I4.0) Concept and Its Tools

In 2011, a team of representatives from various fields and sectors (primarily business, politics, and academia) developed and announced the concept of Industry 4.0 [1]. The basic premise for this concept was the creation of cyber-physical systems (CPS) based on continuous communication via the Internet that allows a constant interaction and exchange of information between humans (C2C), humans and machines (C2M) but also between the machines themselves (M2M) [2]. The announced initiative aimed to stimulate economic development, introduce technological innovation and increase the competitiveness of the German manufacturing industry. For this reason, the German Federal Goverment announced that Industry 4.0 would be an integral part of the "High Technology Strategy for Germany 2020". This further strengthened the interest of the scientific and industrial communities in the new concept and determined its global success and further evaluation.

There are many definitions of Industry 4.0 in the literature. Canas et al. proposed that the I4.0 concept should be seen as "flexible production systems, with or without human interaction, that clone their knowledge and experience with algorithms, models, heuristics, metaheuristics, metaheuristics and hyperheuristics to enable them to be self-configurable, self-maintaining, self-aware, self-designed, sustainable, and both intelligent and smart to make decisions [3]". The concept of Industry 4.0 promotes the development of businesses based on the digitisation and automation of processes and the increasing use of information and communication technology (ITC). I4.0, in its source form, was dedicated to supporting manufacturing companies, as can also be seen in the definition above. The concept was so strongly identified with the manufacturing industry by researchers and managers that phrases such as intelligent manufacturing, smart production or smart industry began to appear as synonyms in the literature [4]. For this reason, in the first phase of the development of I4.0, the main focus of researchers was on digital transformation and automation of production processes [5]. One of the first solutions developed under the I4.0 concept was the so-called Smart Factory, which stood for digital or intelligent factory [6]. The proposed model of the smart factory was based on a fully integrated and robotic production system in which humans should perform only control and monitoring functions, and the information and data needed to perform tasks should be shared, interpreted and exchanged between machines for the efficient execution of manufacturing operations [7]. Therefore, smart factories require manufacturing systems to be equipped with sensors, machines, conveyor belts and robots, the integration of which enables the automatic exchange of information and the intelligence needed to predict and maintain machines.

Based on the smart factory model, many researchers in their publications identify the features characterising Industry 4.0 solutions. Among the most commonly described features are [8, 9]:

- Digitisation, optimisation, and customisation of production.
- Automation and adaptation.
- Human–machine interaction (HMI).
- Value-added services and businesses.
- Automatic data exchange and communication.

The distinguished characteristics of I4.0 are strongly linked to the use of information and communication systems and the construction of complex algorithms responsible for controlling automatic devices. For this reason, some authors identify I4.0 with the industrial process of value addition and knowledge management [10]. In their research, Stock et al. even identified the main trends and expected development of value creation factors associated with implementing Industry 4.0. The results of their study are shown in Fig. 3.1.

The progressive evolution of Industry 4.0 means that the concept now encompasses the digital transformation not only of the industrial market but also of the consumer market, and thus from intelligent production through the entire value delivery channels [11]. At the same time, the fourth revolution is no longer limited to technology alone but is seen much more broadly. Nosalska et al. [12] define Industry 4.0 as a concept referring not only to technological but also organisational changes aimed at integrating the value chain and developing a new business model. These changes are driven by customer needs and requirements for mass personalisation but are made possible by innovative technologies, connectivity and IT integration. Orrt et al. [4] note that I4.0 is bringing about profound technological changes in manufacturing processes, accompanied by organisational changes aimed at optimising costs and increasing efficiency in using company resources. Only an integrated change implementation process will make it possible to increase the flexibility of production processes concerning product/design customisation and agile adaptation of production volumes and schedules to market demand. The results of Veile et al. also confirm this [13] based on an analysis of I4.0 solution implementations in 13 large companies. These authors emphasise that organisational and human aspects are equally crucial during I4.0 implementation in addition to technology and technological processes. Their neglect can delay or completely block the implementation process. Indeed, Industry 4.0 is based on intelligently, horizontally and vertically connecting people, machines, objects and ITC systems, and its impacts are analysed in the four dimensions shown in Fig. 3.2.

Vertical integration refers to the internal processes carried out within a company. It is geared towards digital transformation and intelligent process coordination through

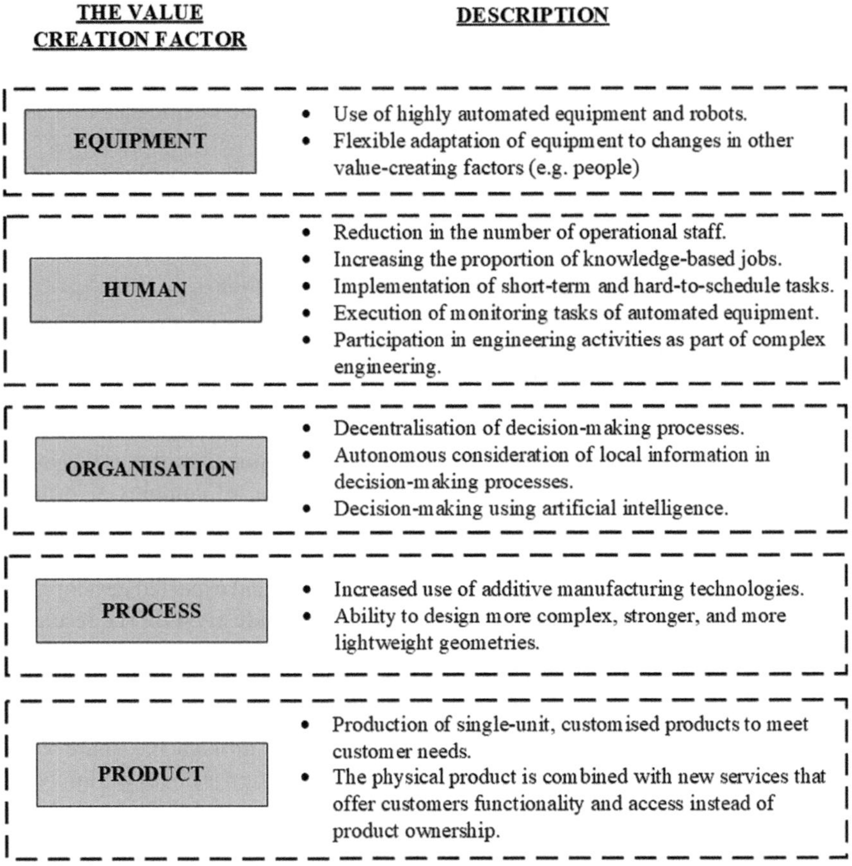

Fig. 3.1 Development of value creation drivers in I4.0 (based on [6])

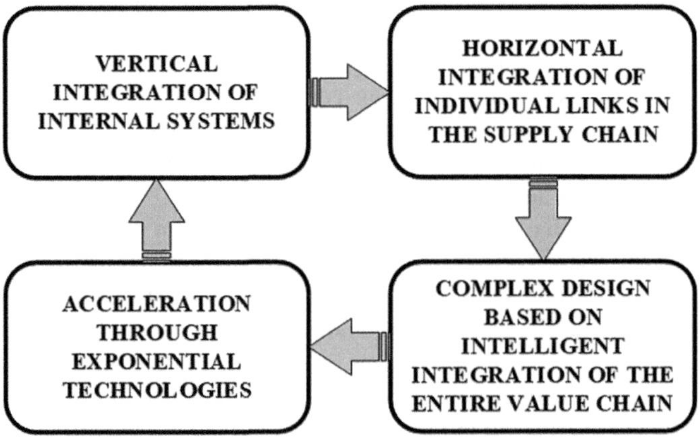

Fig. 3.2 The four dimensions of Industry 4.0 (based on [14])

a cyber-physical production system that considers rapid changes in inventory demand and supply [14]. In this system, the resources used are connected to an integrated network and monitored by smart sensor technology so that data can be collected at the different stages of the processes. Vertical integration is based on the interconnection of internal ITC systems and interfaces and the multilateral exchange of data between the different organisational levels within the company[13]. Until now, departmental employees and the systems they use functioned separately. Thanks to vertical integration supported by I4.0 solutions, these teams virtually combine their activities around the achievement of their goals.

With the implemented technologies of Industry 4.0, it is possible to optimise processes within the value chain of a given company and across the entire supply chain based on vertical integration. This integration refers to linking customer, supplier and service provider systems with the ITC systems of the enterprise [13]. This allows entire supply chains and their customers to be digitally connected in real-time but requires standardisation of interfaces, data types and communication protocols. Despite the challenges that this integration presents, companies are interested in this form of I4.0 implementation because of the achievable benefits [13]:

- Better alignment of supply chain processes.
- Increased efficiency in using material and energy resources while reducing waste.
- Ensuring integrated transparency of partners in the chain.
- Facilitated data exchange and analysis.
- Reduced costs and increased productivity.
- Optimisation of decision-making processes.
- Develop and market highly personalised and innovative products and services.

The third feature is an end-to-end design based on intelligent integration and digitisation used throughout the product lifecycle [14]. Collecting digital data on a product's design and operation process in the different life cycle phases makes implementing learning processes and continuous product improvement possible. Based on the collected data, tailored proactive service strategies are implemented, and the required material structure and manufacturing technology changes are defined. This makes it possible to generate a more flexible manufacturing process that considers changes in customer needs and technological developments.

The use of I4.0 technologies in industrial systems is becoming increasingly popular, and their impact and influence on manufacturing processes are increasing. The industrial space is also seeing more and more solutions using modern manufacturing technologies, big data and networking. Implementing automation, robotics, artificial intelligence, and sensor technologies accelerates manufacturing systems' development, optimising production times and minimising manufacturing costs [14].

Applied technology solutions are a critical component of the Industry 4.0 concept. However, not every ITC-based technology deployed immediately implies an intelligent solution. Qin et al. [15] reviewed the technologies used in I4.0, and according to their research, the intelligence of the solutions used varies. This can form the basis for classifying these technologies by level of intelligence into the three categories shown in Fig. 3.3.

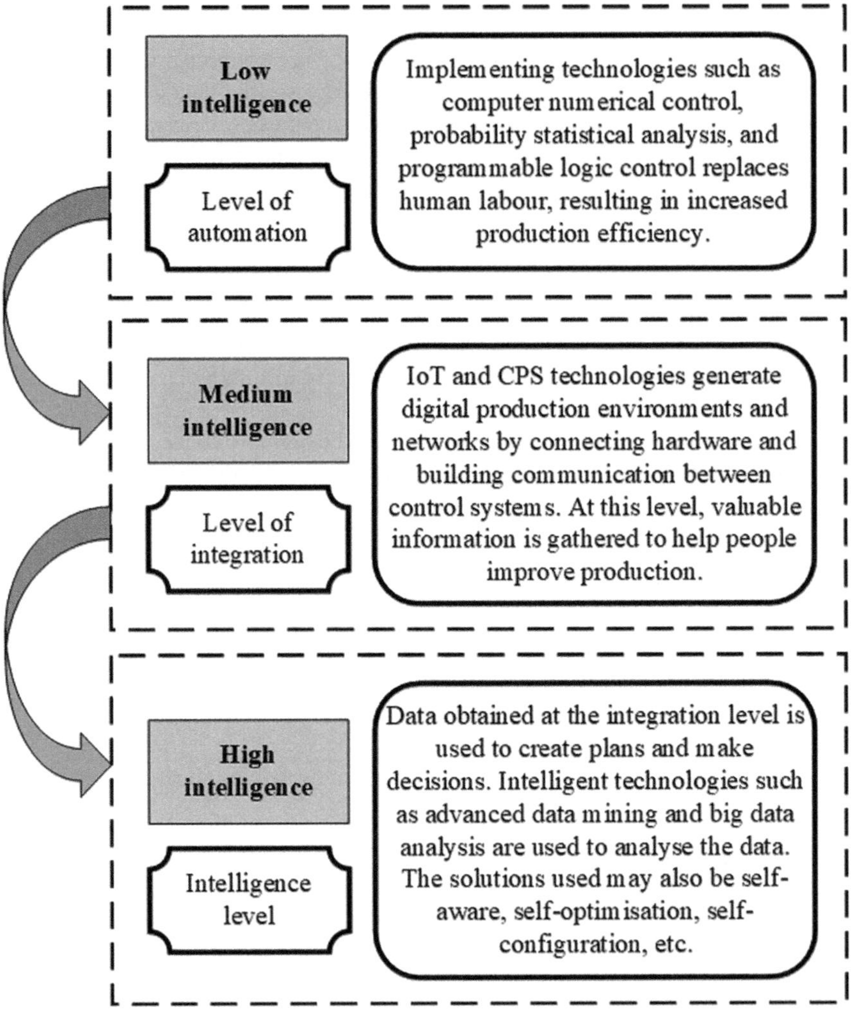

Fig. 3.3 Categorisation of technology by the level of intelligence (based on [15])

An important issue related to the fourth industrial revolution is the principles/ guidelines for implementing Industry 4.0 in companies. For this reason, many authors in their research have attempted to define design and implementation principles for successful I4.0 concepts (including [3, 15–17]). These design principles enable managers to anticipate the progress of I4.0 adaptation in their processes and give them the knowledge to develop the appropriate procedures and solutions required when implementing the fourth industrial revolution [17]. Based on an analysis of 130 articles, Canas et al. [3] identified nine primary principles for designing Industry 4.0 systems described in the literature. These principles are presented in Table 3.1.

Table 3.1 Principles of Industry 4.0 (based on [3, 16])

Principle	Description
Interconnection	The use of technology that enables communication between different electronic devices and between devices and people. Common communication standards that allow flexible linking of machines from different suppliers play an important role
Information transparency	The use of technologies and algorithms to increase the transparency of data exchanged between facilities and feeding decision-making processes. Analysis results must be embedded in supporting systems that are accessible to all participants in the processes to ensure data transparency
Decentralised decision making	Decentralisation of decision-making is possible primarily through CPS and must be supported by automation and real-time data availability. Computers, sensors and actors enable monitoring and autonomous control of the physical world
Technical assistance	Due to the increasing complexity of CPS and the associated decision-making process, supporting systems must support humans. These systems aggregate and visualise information comprehensibly, enabling decision-makers to make informed decisions quickly and solve urgent needs
Human factor	The role of the human being is to make planning and control decisions. However, he or she must meet the competence requirements needed to manage I4.0 systems
Technology	The basis for developing I4.0 solutions is the development of technologies based on digitisation, automation, and ITC systems. Incremental manufacturing is also an important technology related to the operation of smart factories. From the point of view of implementation goals, the compatibility and integration of technological solutions introduced into already functioning manufacturing processes is important
Intelligence	Cyber-physical systems must have a tolerance to the defects that occur through possible self-configuration and restarting, which will increase the overall performance of the machines. Self-awareness of the machines, i.e. the machine's recognition that it is part of a whole and knowledge of its role in achieving the purpose of a given system, is also important
Knowledge	I4.0 systems are designed to collect and provide information about processes and phenomena that will increase the knowledge of decision-makers and the algorithms that control the operation of machines and processes. By acquiring new knowledge, it is possible to increase the flexibility of responses to emerging customer needs and environmental phenomena
Interoperability	This is an important issue for the exchange of information and services. From a technology perspective, interoperability refers to two heterogeneous computing systems' ability to jointly provide reciprocal access to their resources. In relation to supply chains, it refers to the capacity to interact (exchange data and services) among business systems

Veile et al. also analysed 13 implementations of the I4.0 concept in large German manufacturing companies. The results of their research allowed them to formulate six critical conclusions regarding the implementation of the Industry 4.0 concept itself [13]:

- The implementation of I4.0 forces changes in the competencies of employees and their training methods.
- The demand for innovation and new knowledge changes the company's relationship with scientific and research institutions.
- It is necessary to implement organisational changes to flatten and decentralise the decision-making process and establish relations with spin-off companies.
- It is necessary to implement changes in the organisational culture, which should be oriented towards flexibility, openness, a desire for continuous learning and improvement and an entrepreneurial mindset.
- Building relationships based on openness and trust that support collaboration and data sharing with partners across the supply chain is essential.
- I4.0 solutions should be integrated with other machines and production systems.

Other researchers have also confirmed the validity of such implementation conclusions. For example, Bosman et al. [18] cite among the most significant barriers to I4.0 implementation, among others: (a) problems in hiring employees with the right competencies to implement and operate I4.0 solutions and (b) lack of interoperationality of new technologies or their lack of conformance to standards with other technologies in the company.

When implementing the Industry 4.0 concept, it may be necessary to modify the company's organisational structure, which affects organisational culture and decision-making processes. Above all, cyber-physical systems provide the required information for all levels of the organisation, including the operational level, enabling the decentralisation of decision-making processes. Additionally, intelligent factories are characterised by flat, less formal structures, supporting decentralised and optimised decision-making processes [13]. Therefore, agile management methods (including the scrum approach) and a focus on their core competencies are being implemented in these factories.

Digital transformation is also causing changes in employees' work systems. In intelligent factories, people and technology complement each other. Digital support systems aim to relieve the mental and physical strain on workers in the tasks at hand by automating the transport and handling of processes, especially monotonous and repetitive operations [19]. However, there is a real concern that the development of technology will destroy specific jobs, which in the long term may lead to so-called technological unemployment [2]. Digital transformation will also contribute to changing job profiles in many companies. In intelligent factories, there is a change in the role of the human being, who ceases to be a machine operator and becomes a decision-maker and problem solver [16]. Therefore, adaptation measures will be needed to educate and develop employees who need specific digital competencies. In doing so, Hecklau et al. [20] distinguished four types of such competencies: technical, methodological, social and personal. Operating in cyber-physical systems also means that employees must be prepared for regular task assignments and team membership changes. Therefore, they should have the skills to organise themselves flexibly in communities where the critical fit between team members is the competencies required to jointly complete their assigned task within a given timeframe [13]. Critical elements of employee competence development required by Industry 4.0 include [13, 21]:

- Acquiring interdisciplinary knowledge of the processes in which employees participate.
- Understanding the interrelationships that exist between the systems used.
- To have a basic knowledge of ICT systems, automation and data analysis skills.
- To increase awareness of the risks of data misuse and ICT security.
- To build confidence in technology.

The above competencies should be acquired and improved through training and educational programmes, which can also use the effects of digital transformation, e.g. training based on virtual reality solutions. Therefore, cooperation between industry and universities is needed to create innovative technological solutions and analytical methods and develop curricula that align with the requirements of current and future human–machine collaboration environments.

Another critical research area related to implementing Industry 4.0 is analysing its impact on enterprise operations. The I4.0 system design principles outlined above have drastically changed companies' design and delivery of products and services [22, 23]. These changes concern not only the technological solutions used but also new business models affecting the entire value chain. Such models include Manufacturing as a Service (MaaS) and Platform as a Service (PaaS). I4.0 implementations also affect other areas of modern businesses, most notably [9]:

- Large-scale implementation of mass product personalisation through the use of IT.
- Automatic support and flexible adaptation of production systems to changing customer requirements.
- Creating products and components with tracking, communication in their environments and self-awareness.
- Providing new human–machine interaction and operation methods, creating a collaborative robot-human environment.
- Optimising production by improving communication in the smart factory, enabled by, among other things, the Internet of Things.

Based on their research, Pereira and Romero's conclusions indicate that contemporary implementations of I4.0 solutions lead to potentially profound changes in many areas that already go beyond the industrial sector [24]. These authors have identified six critical impact areas for I4.0 systems, shown in Fig. 3.4.

The descriptions above confirm that the ongoing digital transformation and automation significantly change organisations' technical systems, processes and business models. These changes are causing companies to face several challenges in running their businesses and gaining markets in new competitive conditions. Indeed,

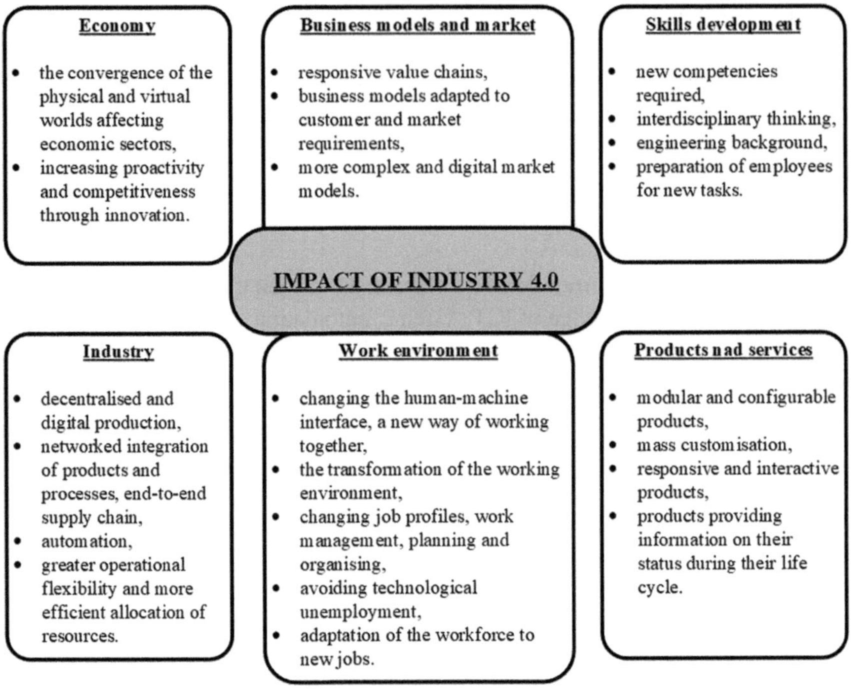

Fig. 3.4 The six areas of influence of Industry 4.0 (based on [24])

as has been highlighted many times, the consequences associated with implementing I4.0 concern technological changes and affect management methods, process coordination, jobs created and eliminated, product life cycles and manufacturing processes. Based on a literature review, Moktadir et al. [14] defined ten challenges to implementing I4.0 solutions in modern companies. These challenges and a brief overview are presented in Fig. 3.5.

From the point of view of the topic of this book, one more aspect related to the implementation of Industry 4.0 deserves attention. Ghobakhloo and Ching [17], in their research, note that implementing I4.0 solutions in risk management processes is multifaceted. The use of IoT, cloud computing, and advanced analytics removes information silos and streamlines information flows within the organisation and across the supply chain. This leads to greater end-to-end (E2E) visibility, which reduces production and consumer risk and improves stability [25]. Tools such as intelligent cameras, smart safety wearables, smart sensors, and AI-based location awareness systems can detect and report dangerous human or machine behaviour [26]. This makes it possible to identify potential hazards in real-time and take targeted action to reduce their impact and likelihood of occurrence. At the same time, many I4.0 technologies have advanced safeguards to enable machines to operate safely and reliably. Increasingly popular are solutions for maintenance management that allow real-time and autonomous asset troubleshooting and problem-solving, which improve the

CHALLENGES	DESCRIPTION
Data security	Companies embarking on an I4.0 implementation lack adequate data protection systems and have developed good practices in cyber security.
High investment costs	Digitisation and automation of operational processes require high investment in developing the technologies used.
Required technological infrastructure	Companies need an adequate technological infrastructure to support the implementation of Industry 4.0 in organisations.
Stable inter-enterprise connectivity	Inter-enterprise communication and real-time data exchange require stable connectivity to implement I4.0 in collaboration.
Workforce reduction	The implementation of automation and robotisation reduces the need for human labour. This creates resistance to implemented changes among workforce team members.
Strategy for Industry 4.0	Industry 4.0 requires implementing dynamic strategic plans that consider digital transformation requirements and changing market demands.
Impact on the environment	Industry 4.0 has a strong impact on the environment - both positively and negatively. Its implementation may reduce the amount of waste generated, but at the same time, the increased demand for energy may cause environmental pollution.
New expectations of management	Managers of Industry 4.0 systems must be newly qualified to implement new and innovative business models based on digitisation and automation.
Need to integrate IT and operational technologies	Digital and physical systems need to be strongly integrated to achieve the full spectrum of benefits of combining the digital and physical worlds.

Fig. 3.5 Challenges of implementing Industry 4.0 in an organisation (based on [14])

security of production environments [27]. At the same time, advances concerning the development of artificial intelligence and machine learning mean that the intelligent collaborative robots (so-called cobots) being developed offer an increasing ability to identify and assess the risk of adverse events. Schou et al. [28] demonstrate that intelligent cobots better interpret the environment, and their use reduces operational risks and provides excellent safety for colleagues.

The concept of Industry 4.0 focuses on the merging of two worlds—the real and the digital, which are supposed to co-exist and, above all, cooperate and support each other. For this reason, the concept of Industry 4.0 is strongly identified with the creation of so-called cyber-physical systems (CPS). These systems are defined as innovative technologies enabling interconnected systems management by integrating their physical and computational environments [29]. Pereira and Romero

[24] emphasise that CPS represent one of the most significant advances in computing and information technology development. These systems create the potential for the interaction between the physical and the virtual environment, integrating, controlling and coordinating processes and simultaneously providing and using data accessing and processing [30]. However, it should be noted that the CP systems described in the literature are characterised by three phases of transformation [3]. The first generation of CPS was primarily based on RFID technology. The data collected in the system came from reading RFID tags and was delivered to the relevant IT systems supporting production and logistics processes. The second generation of CPS already used collections of sensors responsible for collecting the relevant data. The sensors used here, however, had a limited range of available functions. Only the third generation of CPS extensively uses sensor technologies and actuators compatible with networks. This makes it possible to collect, store and analyse data in line with the requirements of today's Industry 4.0 trends.

The continuous development of the Industry 4.0 concept results in an increasing number of solutions identified as I4.0 tools in the research and industrial space. The progressive technological transformation is based on the implementation and integration of simple and advanced information, digital and operational technologies, such as industrial sensors and controllers, Automated Guided Vehicles (AGVs), robots, data analytics, cloud computing, Internet of Services, Internet of Things, Augmented and Virtual Reality (AVR), and Artificial Intelligence. Ghobakhloo and Ching [17] note that many of these technologies have been available to industry for years. However, it is only Industry 4.0 that has brought them to maturity in terms of the integrity and interoperability required in today's digital solutions. Ghobakhloo and Ching [17] also highlight that modern technology trends such as the Internet of Things (IoT) or Cyber-Physical Production Systems (CPPS) are not off-the-shelf technological products but are based on the deployment and integration of various combinations of information, digital and operational technologies into value networks. Indeed, none of the technological trends categorised as I40 solutions work alone. This is why the system interoperability described above, which provides the required level of communication between the elements of an anthropotechnical system and enables data to be shared and coordinated, is so essential [31].

However, the literature review makes it possible to limit these solutions to the eight technological trends that appear most frequently in the results of studies on the I4.0 concept. These solutions are shown in Fig. 3.6.

3.1.1 Big Data Analytics

The creation of cyber-physical systems has resulted in an exponential accumulation of large amounts of data companies use for decision-making, monitoring and control processes and technical systems. Therefore, processes relating to Big Data analysis are linked to Industry 4.0 and the operation of cyber-physical systems.

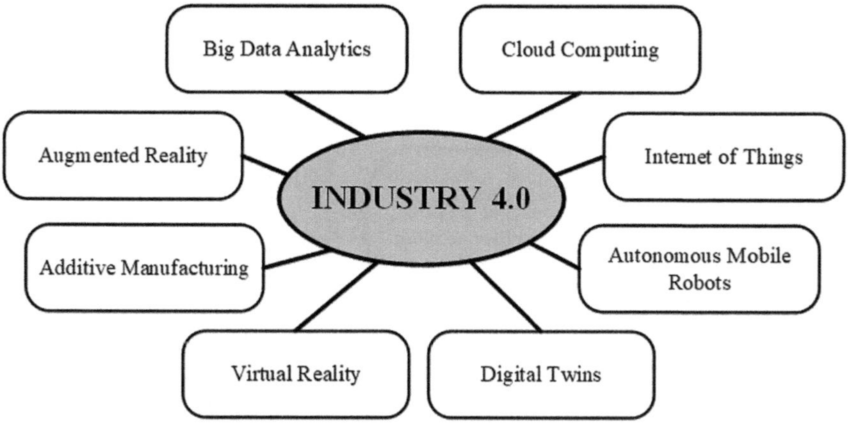

Fig. 3.6 Industry 4.0 solutions

The term Big Data has many different definitions in the literature. Zhang et al. [32] note that the term refers to a dataset of large volume, various categories and complicated structures which need novel frameworks and techniques to excavate helpful information effectively. Elgendy and Elragal [33] highlight that this data's scale, distribution, diversity, and timeliness require the use of new technical architectures, analytics, and tools. Many authors refer to the 5 V model when characterising Big Data collections, as shown in Fig. 3.7.

The diversity of the structure and complexity of Big Data is due to the emergence of new forms of data generated by sensor networks, mobile applications used and various devices that generate and deliver data to technical systems. This has led to an increased demand for the development of tools and analysis methods, considering the complexity of new data sets. This is because access to large amounts of data does not guarantee timely and effective decision-making. The diversified structure and large volume of available data mean that these datasets can no longer be easily analysed based on traditional data management methods and analytical techniques used previously [33]. Therefore, it is necessary to develop new analytical tools and methods for Big Data collection and formulate requirements for system architecture to store and manage such data. Therefore, the required changes related to the management of Big Data relate to three areas: (1) big data storage and architecture, (2) data and analytics processing, and (3) big data analyses.

Elgendy and Elragal [33] proposed a framework for dealing with Big Data analysis for decision-making processes that cover the above areas. The model included various tools for storing, managing, and processing big data, as well as analytical methods and visualisation and evaluation tools used at different stages of the decision-making process. The basic assumptions of the model are shown in Table 3.2.

The availability of big data has significantly influenced the decision-making processes implemented by the company. First and foremost, the datasets created include information from various sources, both internal and external. Companies

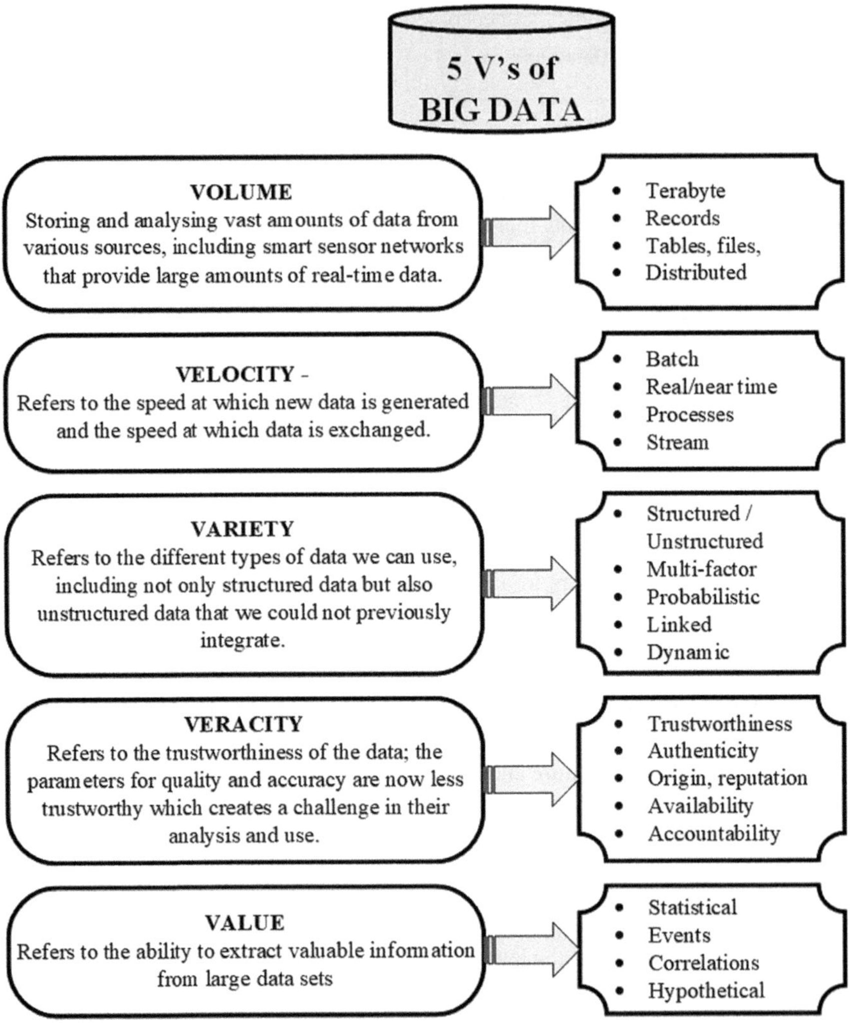

Fig. 3.7 The 5 V model of Big Data (based on [34])

are used to using internal data in their operational planning processes. However, today's decision-making requirements force a strong adaptation of the plans created to external situations that change dynamically and modify the decision-making conditions that must be considered in planning processes. Thanks to the creation of big data, decision-makers have access to relevant information and valuable knowledge, significantly improving the effectiveness of created operational and strategic plans. At the same time, this accessibility makes it possible to create multivariate action plans for different development scenarios of both organisational conditions and market trends. What is particularly important is that the collected data comes from various sources

Table 3.2 Big Data management framework (based on [33])

Area	Description
Big Data storage and management	The big data environment calls for Magnetic, Agile, and Deep (MAD) analysis skills, which differ from a traditional Enterprise Data Warehouse (EDW). They should attract all data sources regardless of quality. Creating an agile database whose logical and physical contents can adapt in sync with rapid data evolution is also necessary. As current data analysis uses complex statistical methods and analysts need to be able to explore data in detail, the Big Data repository must be deep and serve as an algorithmic runtime engine
Big Data analytic processing	There are four critical requirements for Big Data processing: (1) fast data loading, (2) fast query processing, (3) highly efficient utilisation of storage space, and (4) robust adaptivity to highly dynamic workload patterns. Big Data is analysed by different applications, in different ways, by different users with their own objectives. Therefore, the underlying system should be highly adaptive to unexpected dynamics in data processing and not specific to certain workload patterns
Big Data analytics	Data analytics is the process of implementing algorithms to analyse datasets and extract valuable, previously unknown patterns, relationships and information from them. Popular Big Data analysis methods include association rules, clustering, classification and decision trees, regression, and additional analyses such as social network analysis, social media analysis, text mining, opinion mining, advanced data visualisation, and visual discovery

and is collected and available in one place. This positively impacts their comprehensive coverage of decision-maker's information needs and the speed of data distribution for analytical processes. Of course, the ability to provide the expected benefits is strongly linked to the creation of correct mechanisms for accessing and processing the collected data sets.

From the point of view of the subject matter of this book, it is important to note that never before have experts responsible for management processes had access to such large and comprehensive datasets of external and internal phenomena that can feed risk assessment processes, both from the point of view of their probability of occurrence and possible consequences. The wealth of historical data makes it possible to study cause-and-effect relationships, based on which it is possible to analyse the triggers of different types of risks. Real-time data makes it possible to monitor the correctness of implemented processes and assess the effectiveness of implemented changes in risk management. In conclusion, it can be said that the availability of big data increases the potential analytical capabilities at various stages of risk management, from identifying undesirable events through analysing their probability and consequences to monitoring the stages of implemented risk mitigation solutions.

3.1.2 Cloud Computing

As noted above, the idea of Industry 4.0 is based on extensive access to large amounts of data that a company can use to optimise processes, improve products and increase the effectiveness of decision-making processes. Therefore, cloud computing is the primary technology that will enable I4.0 implementation. Canas et al. [3] in their research refer to the definition of cloud computing as an access network demanded by some shared common pool of configurable computing resources (servers, networks, storage, applications and services, among others) that can be quickly provisioned and released with minimum management effort or by interacting with the service provider. The technology is based on scalable IT-related facilities made available to multiple customers as services over the Internet [35]. Customers can use different cloud service models depending on their needs. These solutions are referred to as XaaS, where X stands for any service available through cloud technology, such as hardware, software, databases, platforms or selected types of infrastructure. Three basic service models are available in cloud computing, as shown in Fig. 3.8.

Cloud technology has placed ITC systems in virtual structures, allowing them to be accessed from anywhere at any time. For this reason, Yen et al. [36] stated that the cloud platform is an essential element for the development of CPS. The data storage, processing and sharing capabilities it provides allow the implementation of new business models that take advantage of the availability of resources in the form of a centralised and expandable system. This allows intelligent networking using implemented cyber-physical systems (CPS) that operate decentralised and self-organised [6]. These systems also use collections of sensors to collect data, which

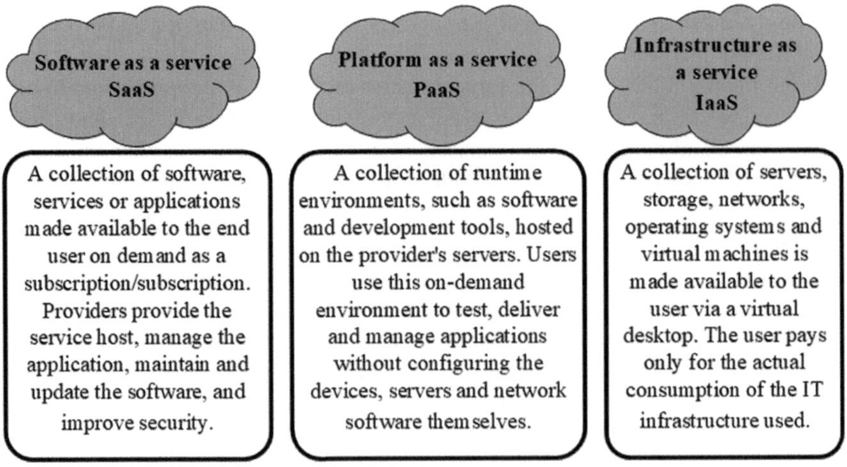

Fig. 3.8 Cloud computing service models (based on [35])

Table 3.3 Advantages and disadvantages of cloud computing (based on [35, 36])

Disadvantages	Advantages
• Lack of transparency in the contracts signed governing access to the service • Data leakage • Need for access to a stable Internet connection • Cancellation of the service provider requires multiple operations to transfer the data to another environment • Choosing a rogue provider may compromise data security and service continuity	• Access to data on different hardware, anywhere, anytime • Reduction of operating costs • Stability and security through advanced security systems • Automatic and regular data backup • Free sharing of data • Uniformity of solutions

are then used to control physical processes. CP systems can exchange real-time data via the Internet of Things through smart connectivity. Based on review publications, the most commonly identified advantages and disadvantages of cloud computing are presented in Table 3.3.

3.1.3 *Internet of Things*

The Internet of Things (IoT) is one of the core tools of Industry 4.0, which is considered a fundamental direction of the fourth industrial revolution [2]. It is defined as a global information society infrastructure that enables the delivery of advanced services through the interconnection of things (physical and virtual) based on existing and evolving information and communication technologies [37]. Although many publications emphasise that IoT aims to offer new services and applications connecting the physical and virtual worlds, where machine-to-machine communication is the primary communication, IoT is increasingly emphasised that it also connects people and machines [37]. This enables the transfer and integration of data between internal business units and across the supply chain. Easier access to information and knowledge makes it possible to increase efficiency in the use of resources, improve the accuracy of planning and monitoring processes, and, most importantly, grow the business based on knowledge management in the organisation.

Since IoT is a network supporting machine-to-machine communication, when discussing this technology, it is worth noting how devices connect and communicate with each other. The literature distinguishes four basic communication models used in IoT solutions, presented in Table 3.4.

It is also worth noting that current IoT solutions are divided into two basic types: Consumer IoT (CIoT) and Industrial IoT (IIoT). CIoT consists primarily of devices that outsource data processing to cloud solutions. At the same time, IIoT typically processes local data supplemented with cloud services to optimise processes and make new services available [38]. The study by Serror et al. identified key differences between consumer and industrial IoT, shown in Table 3.5.

Table 3.4 Communication models for IoT devices (based on [37])

Model	Characteristics
Device-to-device communications	Two or more devices connect directly and communicate with each other without the intermediary of an application server
Device-to-cloud communications	The IoT device connects directly to a web-based cloud service to exchange data and control message traffic
Device-to-gateway model	The IoT device connects via an ALG (application-layer gateway) service, which acts as an intermediary between the device and the cloud service while providing security and other required functions, such as data or protocol translation
Back-end data-sharing model	The communication architecture provides users with solutions to export and analyse smart object data from a cloud service in combination with data from other sources

Table 3.5 Differences between consumer and industrial IoT (based on [38])

Category	Characteristic	Consumer IoT	Industrial IoT
Application	Service model	Human-centred	Machine-centred
	Criticality of the provided services	Not stringent	Mission-critical
Device	Number of devices per home/ factory	Low to medium	Medium to high
	Lifetime	3–5 years	10–30 years
	Hardware complexity	Low	Low to medium
Data traffic	Data volume	Medium	High
	Data confidentiality	Privacy-oriented	Business-oriented
	Traffic type	Periodic and event-driven	Periodic
	Use of communications	Unstructured, contention-based	Structured, planned

When designing solutions using IoT tools, it is necessary to design an appropriate architecture for the system under development. Based on the literature review, four layers of IoT architecture can be distinguished, as shown in Fig. 3.9.

An important issue regarding the application of IoT technology is the digital security of the data exchanged and collected. The deployment of IoT generates new sources of risk that pose a critical security challenge to using this technology. At the same time, users expect assurance that IoT devices and associated data services are adequately protected from potential vulnerabilities and attacks. Improperly secured IoT devices and services can cause cyber-attacks and expose user data to theft. These aspects are increasingly important as technology becomes increasingly ubiquitous in business operations and our everyday private lives [37]. For this reason, implementing measures to address these challenges and ensure the security of IoT solutions must be a fundamental priority for the teams responsible for implementing this technology

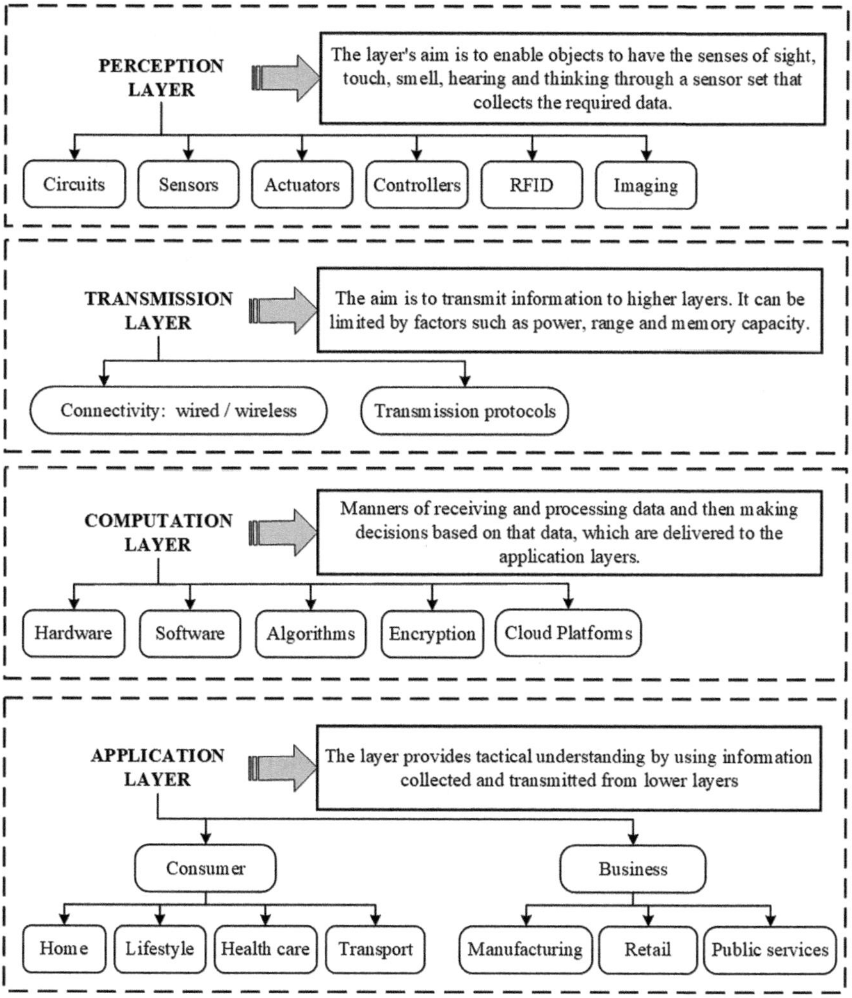

Fig. 3.9 Layers of IoT architecture (based on [39])

into company operations. At the same time, however, it should be noted that often, users themselves are not aware of the risks associated with the use of IoT tools and, in addition, do not know how to configure their networks securely.

3.1.4 Digital Twin

Based on Internet of Things technology, another important tool of Industry 4.0 has emerged—the digital twin. As Vieira et al. [40] note, the digital twin concept is currently interpreted in many ways in the literature and industry, depending on the author and even the sector in which the solution is implemented. While some authors defend that a digital representation of a physical asset or asset system is sufficient to create a digital twin, others argue that a digital twin is much more than a digital representation. Some argue that the digital twin is a technology, and others argue that it is an approach or process rather than a product or technology. For this reason, Kritzinger et al. [41] proposed a classification of digital twins into three types: "digital model", "digital shadow", and "digital twin"—depending on the level of automation of data transfer between physical and virtual resources. The characteristics of this classification are shown in Table 3.6.

Such a wide variety of approaches to solutions referred to as a digital twin led Vieira et al. to develop guidelines that a digital twin should fulfil [40]:

- The digital twin includes a digital representation of the physical asset (asset system) and its context (the accuracy and complexity of the digital representation should match the available resources and purpose of the digital twin being created).
- Each digital twin serves a specific purpose in a given context, thus enabling the definition of the resources needed to support it and the assessment of its benefits and values.
- The digital twin must integrate automated data transfer through sensor monitoring, enabling synchronisation in time between physical and virtual spaces.
- As real-time data alone does not add value to the decision-making process, the digital twin should have some form of data analytics (artificial intelligence, large data sets, etc.) to generate insights for the user (or the twin itself) and support the decision-making process for asset management. Different sources of information, such as physical models and available records, can be integrated into the digital twin so that the twin can use predictive or simulation models, depending on the scope of application adopted.
- Digital twins can have different integration scales, from the level of a single asset or component to the level of an asset system or network. A higher level of asset aggregation in a digital twin means more excellent potential benefits but also greater complexity (data security, interoperability, etc.).
- Digital twins can have varying levels of development and complexity but always involve some kind of automated data transfer—i.e. they take the form of "digital shadows" or "digital twins". The data refresh rate needs to be appropriate for this purpose.

A literature review by Errandonea et al. [42] indicates that the digital twin concept is most often used for maintenance processes, lifecycle optimisation, manufacturing and logic process improvement and design. The application of the digital twin in most publications is described in the context of predicting the condition of an asset to predict an appropriate maintenance plan. These plans are based mainly on forecasting certain phenomena. Digital twins also have applications in preventive strategies, where they are used to predict the condition of an asset to reduce the number of preventive maintenance activities and remove unnecessary maintenance activities

Table 3.6 Classification of digital twins (based on [41])

Type of digital twin	Description
Digital model	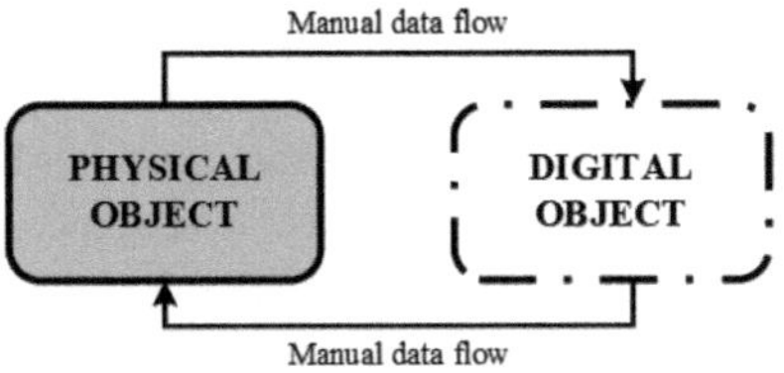
	A digital representation of an existing or planned physical object that does not use any form of automated data exchange between the physical object and the facility (e.g. simulation of the operation of a factory)
Digital shadow	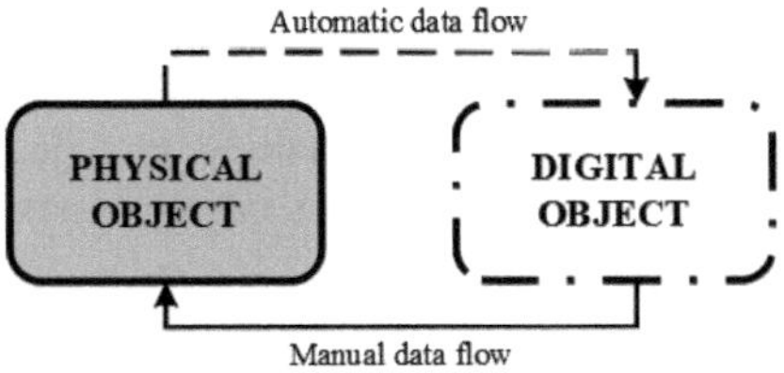
	An automated one-way data flow between the state of the existing physical object and the digital object enhances a digital model. A change in the state of the physical object leads to a change in the state of the digital object, but not vice versa
Digital twin	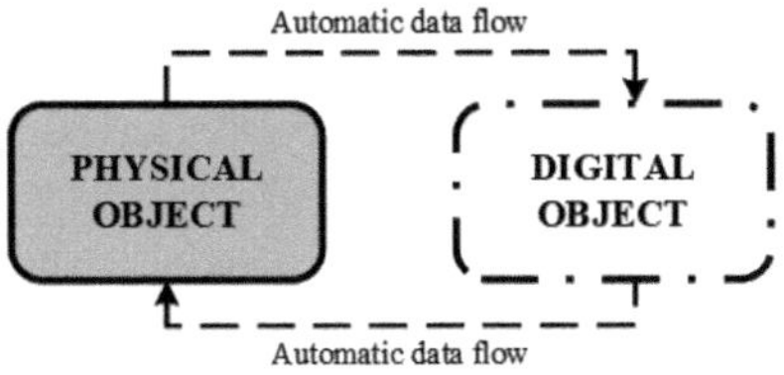
	The data flows between the existing physical and digital objects are fully integrated in both directions. In this combination, the digital object can also act as an instance that controls the physical object. There may also be other physical or digital objects that cause changes in the state of the digital object

by providing longer intervals between them. Digital twins are primarily used in manufacturing and energy sector companies but are also popular in the aerospace, construction, naval engineering, logistics services, rail-way, and automation sectors [42].

3.1.5 Autonomous Mobile Robots (AMR)

Industry 4.0 and the COVID-19 pandemic have sparked a revolution in the use of AMRs, which have become a solution for reducing human–human interaction in favour of human–machine collaboration [43]. Increasing interest in this technology can be seen in academia and industry, resulting in the continued development of AMR. As noted by Rubio et al. [44], robots are nowadays used in very diversified environments, as they support tasks in areas such as surveillance, defence, disaster management, personal assistance, space exploration, petrochemical, industrial automation, construction, marine extreme environments, agriculture, transportation, logistics, sports entertainment, and many other industrial and non-industrial applications. The type of environment in which AMRs are operated influences the way they move. For this reason, Rubino et al. distinguished five categories of robots that differ in their locomotion system [44]:

(1) Stationary (arm/manipulator) robots—the robot's base is stationary and consists of an open kinematic chain, terminated by a specialised tool dedicated to performing specific operations. These include various types of manipulators, grasping devices, and industrial robots.
(2) Land-based robots:
    - Wheeled mobile robots (WMR) are the most crucial robot locomotion systems, and they are used in transportation and logistics. It can use different numbers and types of wheels, which is important for modelling the kinematics and dynamics of its movement.
    - Walking or legged mobile robots—although more expensive than WMR, in their favour is that this type of locomotion is more mobile, has better energy efficiency, stability and less impact on the ground.
    - Tracked robots—have much larger ground contact patches, which significantly improve their manoeuvrability on loose surfaces compared to conventional wheeled robots.
    - Hybrid robots—their design combines any locomotion systems outlined in a–c. There are typically four categories of hybrid mobile robots: (a) leg-wheel locomotion systems, (b) leg-track locomotion systems, (c) wheel-track locomotion systems, and (d) leg-wheel-track locomotion systems.
(3) Air-based robots—commonly called drones or unmanned aerial robots. They perform pre-programmed tasks with or without human intervention.
(4) Water-based robots—underwater vehicle-manipulator systems used to explore the seabed, but can also be used to rescue specific objects (including human beings) by being equipped with, for example, a robotic arm.
(5) Other robots:
    - Snake-like robots—highly articulated robots that can perform tasks in a versatile manner that are not bound to climbing, crawling, and swimming.

- Worm-like robots—move similarly to earthworms, which is highly desirable, particularly in confined spaces.
- Nanorobots—their components are at the scale of a nanometer and allow precise interaction with nanoscale objects.
- Cooperative robotics—an assembly of robots controlled in a coordinated manner.

The locomotion system of a mobile robot is a critical component of its design. It depends on the robot's environment and the required technical criteria such as efficiency, manoeuvrability, or stability [44].

A critical aspect of the application of autonomous mobile robots is their navigation in anthropotechnical environments. Therefore, among the most critical challenges currently facing the design of AMR systems are [45]:

- Robot perception. Robots must have accurate, reliable, and robust perception capabilities to operate effectively outdoors. For this to be possible, they need to be equipped with various sensory sensors that monitor sound, pressure, temperature, light and contact with another object to improve their perception of the environment.
- Localisation and mapping. Based on the information retrieved from the perceptual sensors, the robot's location is determined, and its environment is mapped. AMR requires very accurate and reliable environmental mappings and localisation at a sophisticated level based on the context of the application. The Simultaneous Localisation and Mapping (SLAM) approach is a technique that represents the robot's positions and the map of its external environments. Most SLAM systems rely on LiDAR and vision-based sensors such as Red–Green–Blue/Red–Green–Blue-Depth cameras.
- Scene understanding goes far beyond object detection and requires analysis of the data obtained by the sensors. It is a complex problem that requires more steps than just recording and extracting features. The robot must perform tasks such as depth estimation, scene categorisation, object detection, object tracking, and event categorisation.

Planning the paths of their movement is also an important issue related to the execution of tasks by AMRs. This planning involves identifying the best paths for the robot to move along, taking into account the times of the operations and ensuring that it reaches its destination without collisions. The second essential element related to the operation of AMRs is the planning of their trajectories, which consists of finding out the force inputs (control u(t)) to move the actuators so that the robot follows a trajectory q(t) that enables it to go from the initial configuration to the final one while avoiding obstacles [44]. The robot's dynamics and physical properties must be taken into account when planning the robot's trajectory. Path and trajectory planning are elements of robot motion planning that determine their application's efficiency, effectiveness, and safety.

3.1.6 Additive Manufacturing (AM)

Incremental manufacturing is also known as rapid prototyping, freeform fabrication, or 3D printing. 3D printing derives from the fact that the process of joining materials to make objects is based on 3D model data [46]. Incremental technology is a method of manufacturing components (products) by depositing material layer by layer on or without a substrate [47]. Information on the parameters describing the component is prepared as a blueprint in computer-aided design (CAD) software and then converted to a stereo-lithography (STL) file, which stores data on each layer. A product diagram made in CAD is approximated by triangles and sliced, containing the information on each layer that will be printed [48].

Additive manufacturing is not a very new technology. It was implemented in the industry in the '80s but was initially limited to manufacturing or prototyping small products [49]. However, the fourth industrial revolution stimulated intensive development of this manufacturing technology and opened up new dimensions of engineering applications for it in various sectors of the economy. AM has gained popularity primarily because it addresses selected limitations in conventional manufacturing. Studies reported in the literature indicate that implementing AM processes reduces the need for tools and increases flexibility in designing products and components and adapting them to changing customer needs [50]. It is worth noting here that achieving increased flexibility is possible through using different software and technological platforms, enhanced computational power and connectivity [49]. It also provides improved design accuracy, flexibility and reliability. At the same time, compared to traditional machining, incremental manufacturing enables greater material utilisation (reduced waste), reduced buy-to-fly ratio, reduced machining and manufacturing costs, improved mechanical properties and shorter overall process lead times are also possible [47]. AM offers an unprecedented opportunity to digitise manufacturing processes. Traditional manufacturing procedures require an in-depth analysis of the component geometry to determine the best manufacturing sequence for each component, the equipment needed and any extra fixtures needed to finish the component [46]. AM, on the other hand, requires basic dimensional information, an understanding of the operation of the AM machine and knowledge of the materials used for manufacturing.

The continuous development of AM technology and, above all, the benefits associated with its use indicated above mean that different types of solutions are now distinguished, which are usually categorised according to three subdivision criteria[49]: (a) type of base material used; (b) medium used for processing; (c) methodology of formation of the product. The types of solutions used in each category are shown in Fig. 3.10.

The superiority of AM technology over traditional manufacturing for selected products means that new methods and application areas are constantly being observed, dynamically driving these solutions' market development. The main target areas for the application of AM technology have been the aerospace and automotive sectors, where components with less weight but high strength are expected [47].

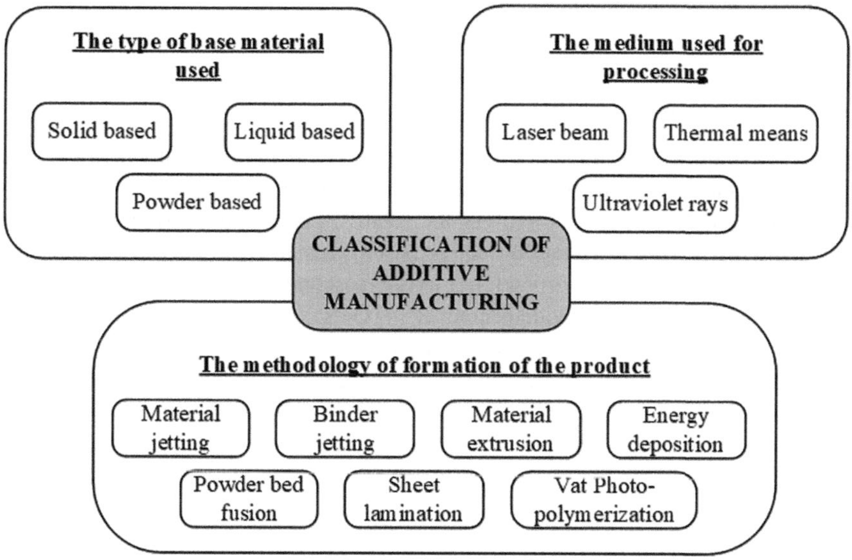

Fig. 3.10 Classification of additive manufacturing (based on [49])

However, we can observe its increasing use and growing popularity in electronics, biomedical, medical, consumer goods, footwear, electrochemistry, chemistry, and many others. Thanks to the introduction of fibre-reinforcement in 3D printed plastics/resin products to improve their mechanical performances has augmented AM's expansion in polymer composite manufacturing, which is used in many sectors not mentioned above[49].

3.1.7 Augmented Reality (AR)

Augmented reality technology is a human-assisted solution to eliminate potential errors made at the operational level. For Masoni et al. [51], AR is the leading technology for I4.0 [3]. This technology connects the virtual (computer-generated) world with the real world, showing humans their simultaneous integration. This means that AR tools provide an interface linking digital information to the real physical system [52]. Therefore, an AR system is an emerging technology with which a person can see more than others see, hear more than others hear, and perhaps even touch, smell and taste things that others cannot [53]. Also interesting is the definition of Daponte et al. [54], who stated that AR enriches the user's sensory perception by showing information about the surrounding environment (e.g. physical quantities) that cannot be perceived with the five senses.

AR is a technology that belongs to the group of so-called mixed reality, meaning that it is a form that combines a virtual and a physical environment [53]. As such, it fulfils three characteristics of this group of systems, namely (a) combines the real-world and virtual objects, (b) runs in real-time, and (c) allows the interaction between the users and virtual objects [55]. To provide this interaction to the worker, AR systems, in their basic structure, consist of the following elements [54]:

- A video camera transmits an image of the actual environment in which the user is located and performs the tasks assigned to him/her.
- Tracking Module, which monitors the relative position and orientation of the camera in real-time. This module can be based on every type of sensor technology: (1) 9D IMU (3-axis accelerometer, 3-axis gyroscope, and 3-axis magnetometer), (2) ultrasonic sensors, (3) video cameras, (4) GPS modules, and (5) RFID devices.
- Graphic Processing Module, responsible for processing recorded images and their integration with virtual objects.
- Display, which is responsible for providing the user with an integrated image of the real world in combination with virtual objects.

Visualisation technology is an important element that determines, among other things, the comfort of working with AR tools. Four such technologies are generally available on the market [56]: head-mounted displays (HMDs), handheld devices (HHDs), static screens, and projectors. These technologies are implemented in AR tools, which can be stationary or mobile devices, depending on the defined visualisation task and the employee's need to move around as part of their duties. In [57], authors point out that most AR solutions involve several devices, as many implementations render in a fixed computer and synchronise the real and virtual objects in a mobile device. However, research presented by numerous authors [56, 57] indicates that HMDs-based solutions are the most widely used in both industry and science. This is because their use does not restrict the operator's movement but allows hands-free access and information reading, significantly speeding up operations [58]. A significant improvement that has further positively influenced the use of this technology is the emergence of AR smart glasses (ARSGs), which show good potential for industrial applications [59]. The ARSGs currently available are equipped with numerous operator-assistive functions, including displaying information to track, distribute, and store data about the surrounding environment and the user [60]. The high functionality of the available ARSG solutions positively influences their increasing popularity and application in industrial processes.

Numerous publications can be found in the literature on the use of AR technology to support manufacturing processes [61], maintenance [62], and internal logistics systems [63, 64]. The implementation of AR tools is often concerned with supporting workers in operations such as [57]: manual assembly, robot programming and operations, process simulation, process monitoring, picking process, maintenance, training, quality inspection, operational setup ergonomics and safety. In addition, it should be noted that the most spectacular results are achieved for processes where, thanks to AR, it is possible to increase productivity by:

- Eliminating or reducing the human errors [58, 65].

- Increasing the flexible delivery of operations to the operator without the operator having to use their hands [66, 67].

The growing popularity of AR tools makes it possible to formulate many benefits of applying this technology to enterprise operating systems and many associated challenges. Figure 3.11 shows the most commonly cited benefits and challenges of implementing AR technology in the industry.

Despite the numerous benefits of AR tools in employee operational support processes, ongoing research indicates that the technology is also associated with some risks. An analysis of the publications presented in [69] shows that most authors classify the potential risks associated with using AR into three categories: technical, organisational and ergonomic. A more elaborate classification of constraints was formulated by de Souza Cardo-so et al. [57], who distinguished five categories concerning users' health and acceptance, tracking methods, projection quality, accuracy and interaction, hardware, and development complexity. It is also worth noting

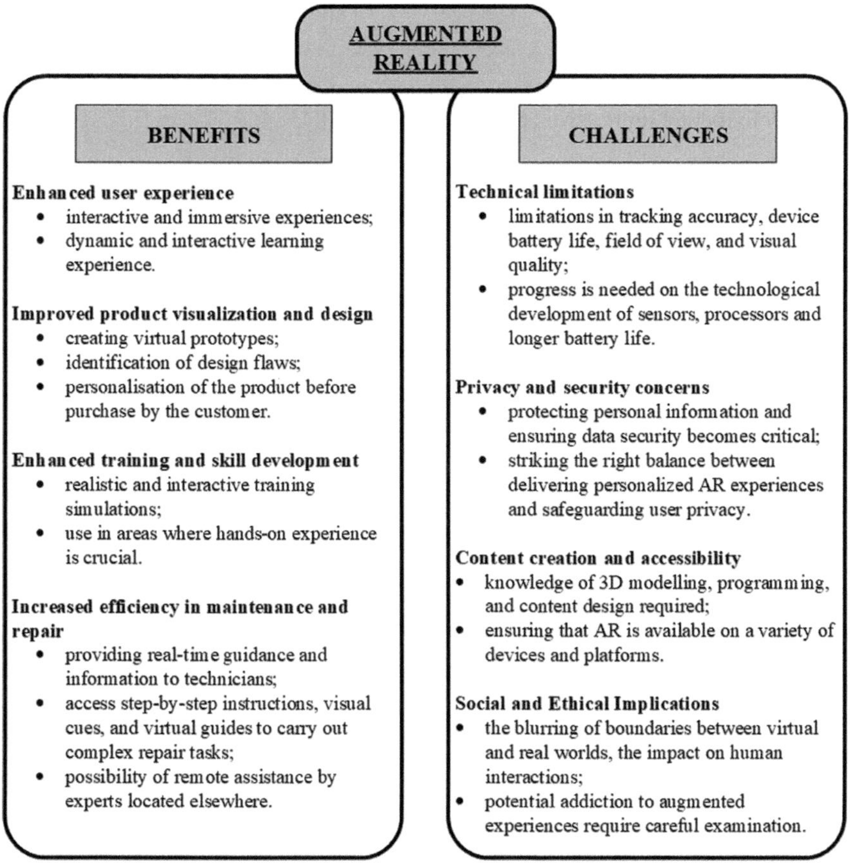

Fig. 3.11 Benefits and challenges of using AR technology in the industry (based on [68])

that the identified barriers to AR implementation occur. Stoltz et al. [70] divided these into four groups: hardware limitations, software challenges, acceptance, and cost.

3.1.8 Virtual Reality (VR)

Virtual reality is also used in processes to improve employee competence. The basis of virtual reality is the artificial creation of a digital environment whose task is to replicate a real, physical system or process as faithfully as possible. Three attributes should characterise the digital environment created:

- immersion—an objective level of sensory fidelity a VR system provides [71];
- presence—the user's subjective experience resulting from being in the immersive environment [72];
- interactivity—the degree to which a user can modify the VR environment in real-time [73].

There are three groups of VR environments, which are divided based on the level of interactivity and immersive environment [74, 75]:

- non-immersive virtual environment (Desktop VR), where conventional PCs are used for training scenarios and keyboard, mouse, wand, joystick or touchscreen are used for exploration—the simplest type of virtual reality applications;
- immersive virtual environment (IVR) uses a multiple, room-size screen or a stereoscopic, head-mounted display unit. Haptic, audio and sensory interfaces may enhance these systems;
- semi-immersive environment (Fish Tank VR)—use of conventional monitors, with simultaneous use of systems to support head tracking and therefore improve the feeling "of being there" thanks to the motion parallax effect.

Training scenarios implemented in an immersive virtual environment are the most effective. The high concentration level on the training tasks due to the high representation of the real system and the lack of distractions for the trainee translates into greater effectiveness of this skill acquisition and development. Of course, the critical element, in this case, is the faithful reproduction of the real system, as any distractions occurring in the scenarios cause the trainee to become distracted, reduce the comfort of the training tasks and, above all, can lead to the occurrence of so-called simulation sickness.

The growing interest in using VR technology in training scenarios is a response to the limitations of traditional learning forms. The most essential benefits described in the literature and the most popular application areas for VR tools are shown in Fig. 3.12.

Many authors emphasise that the available VR tools meet the expectations of today's trainees and trainers, who demand high performance from training systems based on [77]: focus, meaningful representation of information, multiple mappings

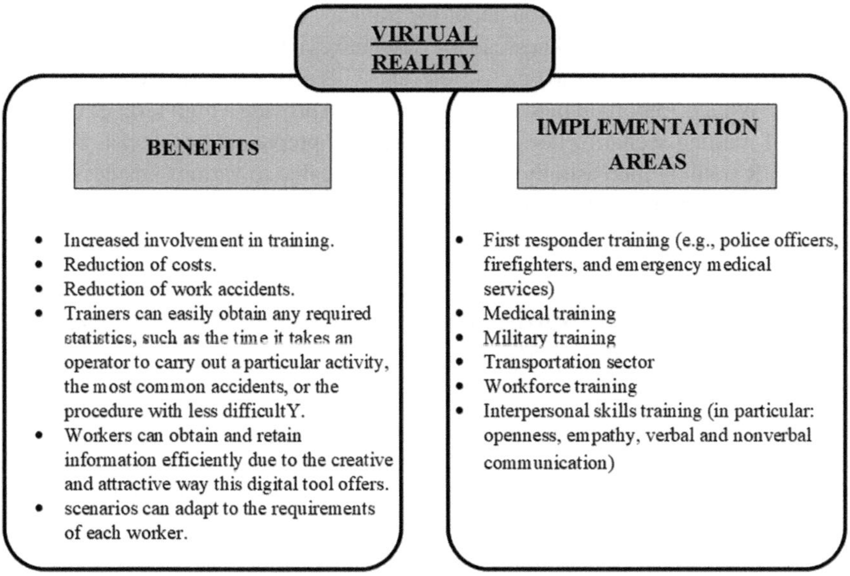

Fig. 3.12 Benefits and areas of application of VR technology in training systems (based on [76])

of information, and reflective learning. The widespread use of VR technology is primarily observed in training, where participation in traditional training involves risking the health or even life of trainees or other participants in the training process. Using VR tools in the first training phase, where trainees still make numerous mistakes and do not develop the right behaviours, is more effective and, above all, safer. For this reason, VR training is widely used to improve the skills of sappers, surgeons, emergency services and many others in which the trainee is exposed to dangerous situations. Therefore, in the early stages of development, VR technology was primarily used to improve competencies for which training processes were associated with risks to the health or life of trainees or very high implementation costs [78]. Researchers on this topic also point out that contemporary training systems should use VR tools, as they enable [79]:

- the use of a large number of training scenarios while reducing implementation costs. Developers create different scenarios based on existing 3D assets (both self-created and acquired), which can be reused many times in different training configurations;
- delivery of training in the privacy of one's own space, which is especially important in training where participants may feel uncomfortable being observed by others;
- the delivery of training under the supervision of an instructor who can react quickly to emerging early feedback on even partial performance of the trainee and counteract any problems or negative trends in his or her performance.

With anticipatory information about the trainee's errors during the training, it is possible to identify their weaknesses [80]. This makes it possible to adapt the learning parameters to allow trainees to practise relevant skills in purposefully adapted training systems. What is important here is that the ability to provide a high level of customisation of training scenarios does not generate high preparation costs [81, 82]. This makes VR training increasingly popular and applicable to various business sectors and processes, including logistics.

3.2 The Impact of Digitisation on the Enterprise's Logistics Processes and Supply Chain

As part of the description of the phenomena influencing the growing interest in the concept of risk in organisations, Chap. 1 presented a general characterisation of the digital transformation phenomenon observed for several decades. It focuses on how digital transformation affects analytical capabilities and the choice of risk assessment and management tools, as well as on emerging risks resulting from the increasing accessibility of the digital organisation. The introduction also introduces the phenomenon of risk intelligence, which will be relevant to the aspects discussed in this book section and defines an organisation's ability to assess risk based on the digital data and experience it collects.

In the literature related to digital transformation research, one encounters two basic terms relating to this trend, namely digitisation and digitalisation. Some authors even make the mistake of using the two terms interchangeably, which requires them to be clearly defined before proceeding further:

- Digitisation is the original term for converting analogue or physical information into a digital format [83]. It is the simplest form of transformation involving converting data from hard copy to digital.
- Digitisation is an extension of digitisation and refers to how social and economic life areas are restructured around communication infrastructure and digital technologies [84].
- Digital transformation (DT) is a phenomenon that has developed from both of the above trends [83], as shown in Fig. 3.13. DT refers to business processes, organisational culture management styles, corporate strategy, and their combination with digital technologies [85]. They aim to transform a company's processes to improve performance.

Currently, there is no clear definition describing the phenomenon of digital transformation. A review of the literature and a critical analysis of the definitions formulated by different authors shows that [85]:

- Authors in the literature present different perspectives on the phenomenon of digital transformation. Some consider it an action strategy, others consider it a process, and some define digital transformation as some disruption.

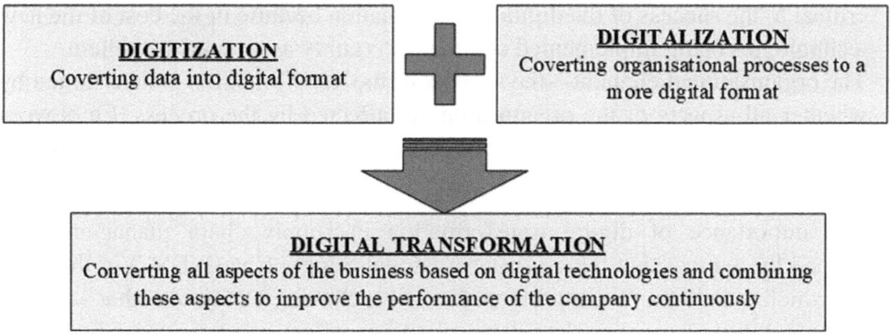

Fig. 3.13 Correlation between digital terms (based on [85])

- A critical element of digital transformation is technology, which is expressed, among other things, in that in most definitions, there is a reference to the concept of "digital technology".
- Some definitions have some ambiguity, such as referring to social welfare or using specific examples, significantly weakening the formulated definition.

Based on the research presented in [85], in this book, the Author will refer to digital transformation as customer-oriented changes and improvement mechanisms that enable continuous productivity improvements in production processes and enhancement of supporting processes through the implementation of advanced digital technologies. The critical point here is that these mechanisms are implemented at the level of the company's internal processes. However, their use may also apply to cooperation with selected partners in the supply chain. In fact, digital transformation is recognised as a critical trend in intra-organisational change, as well as improving cooperation between supply chain partners. These improvements are made possible primarily by ensuring better partner communication, greater transparency and accelerating data exchange processes. Therefore, this section focuses on digitalisation's impact on managing supply chains and logistics processes within a company's internal material handling. Additionally, Abduallah et al. in their research highlight four critical elements associated with such a formulated definition of digital transformation [85]:

- The element of customer orientation—listening to customers' needs and aligning digitisation processes with their expectations is a critical success factor. It will only be possible to achieve the required profitability through digital development if the appropriate level of customer satisfaction is achieved.
- An element of continuity—digital transformation should be a process that never ends. Organisations should continuously transform and improve their processes in line with technological advances and changes in digital customer needs.
- Digital technology element—selecting appropriate digital technologies should consider the specifics of the organisation's operations. At the same time, it is

critical to the success of the digital transformation because of the cost of the new technologies being implemented and the uncertainty associated with them.
- The organisational element—the success of the transformation is determined by whether all aspects of the organisation are affected by the process. Employees must also use digital technologies effectively and efficiently to improve business and supply chain processes.

The importance of digital transformation in supply chain management is confirmed by the fact that there is a term Digital Supply Chain (DSC) in the literature, which stands for "an intelligent best-fit technological system that is based on the capability of massive data disposal and excellent cooperation and communication for digital software, hardware, and networks to support and synchronise interaction between organisations by making services more accessible, valuable and affordable with consistent, agile and effective outcomes" [86]. DSC management aims to achieve the eleven core features shown in Fig. 3.14.

The literature review presented in [87] on the digitisation of logistics and supply chain processes indicates that this topic is widely covered in the global literature. The number of issues raised about this topic is so numerous that it could be material for a separate book. Therefore, for the discussion in this book, the scope of the issues analysed concerning digital transformation in supply chains has been limited to three thematic areas concerning (1) the conceptual framework of DT implementation, (2) the impact of DT implementation on the logistics processes of supply chain participants, and (3) the challenges of this implementation.

3.2.1 Conceptual Framework for Implementation

When analysing the phenomenon of digital transformation occurring both in internal processes of organisations and in entire supply chains, it is first necessary to identify the conceptual framework for implementing digital solutions. Abdallah et al. [88] based the digital transformation implementation model on four pillars: people, enabling technology and tools, business processes, strategy and leadership. The findings of other authors also confirm the validity of this approach. Therefore, it is worth considering the implementation framework, which is thus defined based on selected publications.

(1) People

A critical element in implementing digital transformation is mainly the people affected by this change, who must accept it and be actively involved in its implementation. The importance of an active role in implementing digital transformation is highlighted in their research by Genzorowa et al. [89]. The digital solutions being implemented are repeatedly associated with the need for employees to complete specific knowledge and skills, particularly the development of so-called digital competencies. For this reason, as management embarks on a transformation

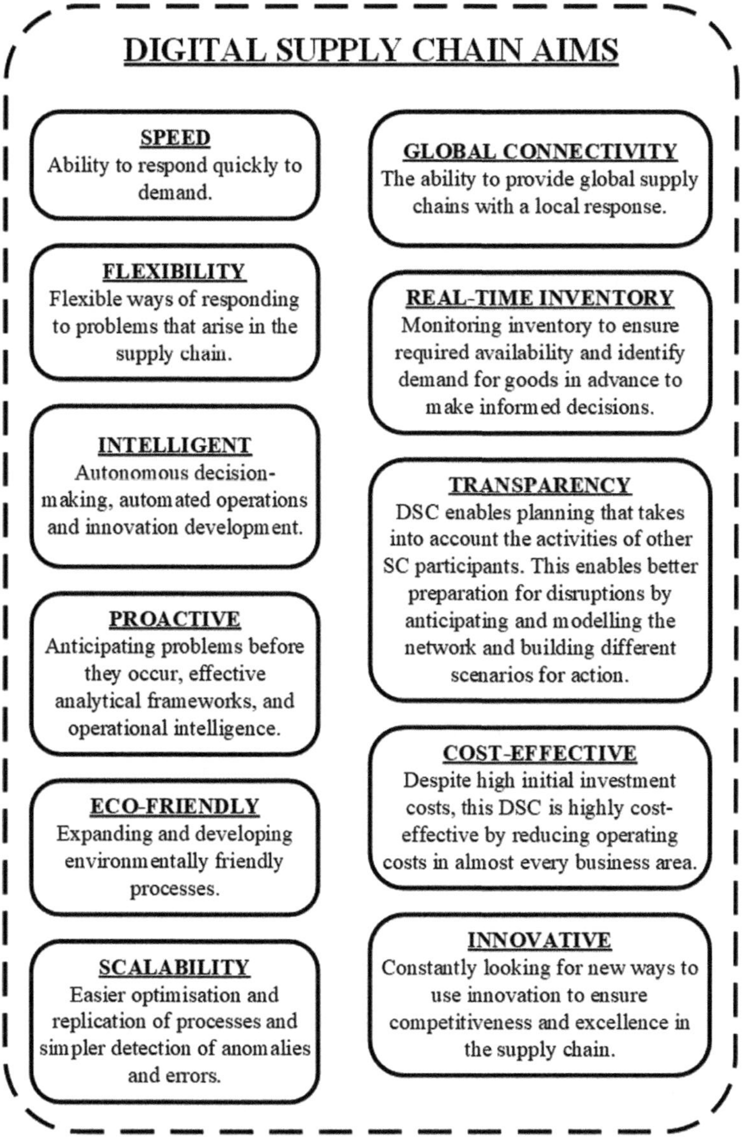

Fig. 3.14 DSC characteristics (based on [86])

initiative, it is necessary to develop individual training plans to enable employees to acquire the required skills and find their way in the new conditions of their tasks. Of course, for this to be possible and effective, it is first necessary to identify existing skills gaps in employees and establish a team to support the implementation of digital transformation at the individual company level. Such a team should include [88]:

- the leader, whose job it is to control the entire process and solve any problems that arise;
- the change agent, who is responsible for motivating people to actively participate in the transformation and flexibly adapt to the new working methods and conditions;
- the business specialist with experience in marketing and sales;
- the data architect, whose job is to analyse and create reports to support the decision-making processes of senior management;
- the financial analyst, who is responsible for preparing cost–benefit analyses and managing the project budget;
- the UX specialist, who is responsible for compliance with implemented changes to the needs and expectations of customers and users.

(2) Enabling technology and tools

Digital transformation is highly dependent on implementing various technological solutions to support the implementation of core and enabling processes within an organisation. The development of the Industry 4.0 concept and associated technologies has provided companies access to a wide range of digital tools and automated and autonomous devices that meet the company's needs at different stages of digital transformation. Thus, depending on their level of digital maturity, individual supply chain participants can choose appropriate technological solutions that are adequate to the digital maturity of their internal processes but also support their digital integration with their supply chain partners [90]. Indeed, as noted by Mittal et al. [91], each organisation should choose the right mix of digital technologies in line with its needs and financial capabilities.

(3) Business processes

Digital transformation often changes business models within a company [89], resulting in the need to redesign individual processes and implement organisational changes. Thanks to the development of digital technologies, new opportunities arise in the coordination of processes in entire supply chains based, among other things, on new standards for data exchange [92], as well as the modernisation of processes within the company, but also in its cooperation with cooperators [93].

A critical issue regarding business processes is integrating and coordinating them along the value chain. This has important implications for the implementation of digital technologies. The application of specific technological solutions needs to be considered and coordinated at the level of the interconnected processes so that the effects achieved are global and do not lead to sub-optimisation of parts of the business. Therefore, the digital development of a company should be based on solutions implemented at the level of individual processes (e.g., manufacturing and logistics) and encompass the entire organisation's activities [90].

(4) Strategy and leadership

Digital transformation is a continuous change process that needs to be integrated into the corporate and supply chain integration strategies. Including aspects of digital transformation in the organisation's defined strategic objectives gives due prominence to the changes being implemented and becomes a critical aspect of the business plans being developed. Setting digital strategic objectives is crucial in determining what an organisation wants to achieve with its digital strategy and separating the important from the unimportant [86]. At the same time, by embedding this development trend in strategic activities, top management expresses its acceptance of the transformation and, therefore, should support the development of an organisational culture that promotes change regarding digitalisation and creates incentive systems that encourage employees to participate actively [88]. Integrating digitally-minded individuals and organisations will enable the continuous development of the supply chain and the collection of shared digital information for improvement.

Including digital transformation in the prepared strategic plans also implies the need to create a strategy for the organisation that aligns with the approach taken to digitisation at the supply chain level. Lipsmeier et al. [94] note that the absence of a digital transformation strategy means that a company will only undertake isolated, small-scale projects whose effectiveness will be low across the organisation and which will waste the resources invested in them. For this reason, many authors focus on the search for effective digitisation strategies that will enable organisations and entire supply chains to achieve rapid and effective digital transformation for different levels of management. This strategy should be a comprehensive vision for the digital development of the enterprise, which describes the goals and tools for its implementation, as well as the value creation for the organisation and the entire supply chain [94]. Measures aimed at aligning the impact of digital technologies with the company's processes and linking them to supply chain participants' internal and network procedures are also important elements of this strategy [95].

Ho et al. [96] found that more than one-size-fits-all approaches to developing a digitisation strategy can be identified. Therefore, they distinguished three main typologies of digitisation strategies for collaborative supply chain organisations: top-down, bottom-up, and mixed. At the same time, these authors confirmed that the choice of digitisation strategy is critically influenced by three factors [96]: (a) the number of links at each level of the logistics network, (b) the market demand reported by customers, and (c) the type of product delivered to the consumer market (divided into standard and specialised products).

The effectiveness of implementing a digital strategy is also influenced by the leadership style adopted and the role of the leader responsible for the process. Abduallah et al. [88] emphasise in their research that a leader should possess the following characteristics: flexibility of thought and openness to change; diverse knowledge and curiosity to learn about other solutions; focus on priorities and results; and accountability for his or her own and his or her team's actions.

The digital implementation framework is also explored in the context of developing different management strategies that support the functioning of supply chains. An example is the findings of a study by Ehie et al. [97], in which the authors proposed using digital technologies to improve core supply chain management processes within the SCOR strategy. This research aimed to understand the conditions under which supply chain digitisation is more or less effective in improving the integrated processes of planning, purchasing, production, distribution and returns.

Based on their research, Abduallah et al. [88] also formulated rules for implementing digital transformation in organisations, which can constitute a set of good practices for managers responsible for this process. These rules are shown in Fig. 3.15.

The good practices outlined above are universal. However, when embarking on implementation, we must consider the specificities of the company in question due to its role in the supply chain and the sector it serves. For this reason, numerous publications in the literature are narrowed down to the specifics of a particular sector served by the supply chain, for which researchers attempt to identify rules for DT implementation that consider the requirements of a specific group of customers. Examples of such publications are characterised in [98], among others. The results

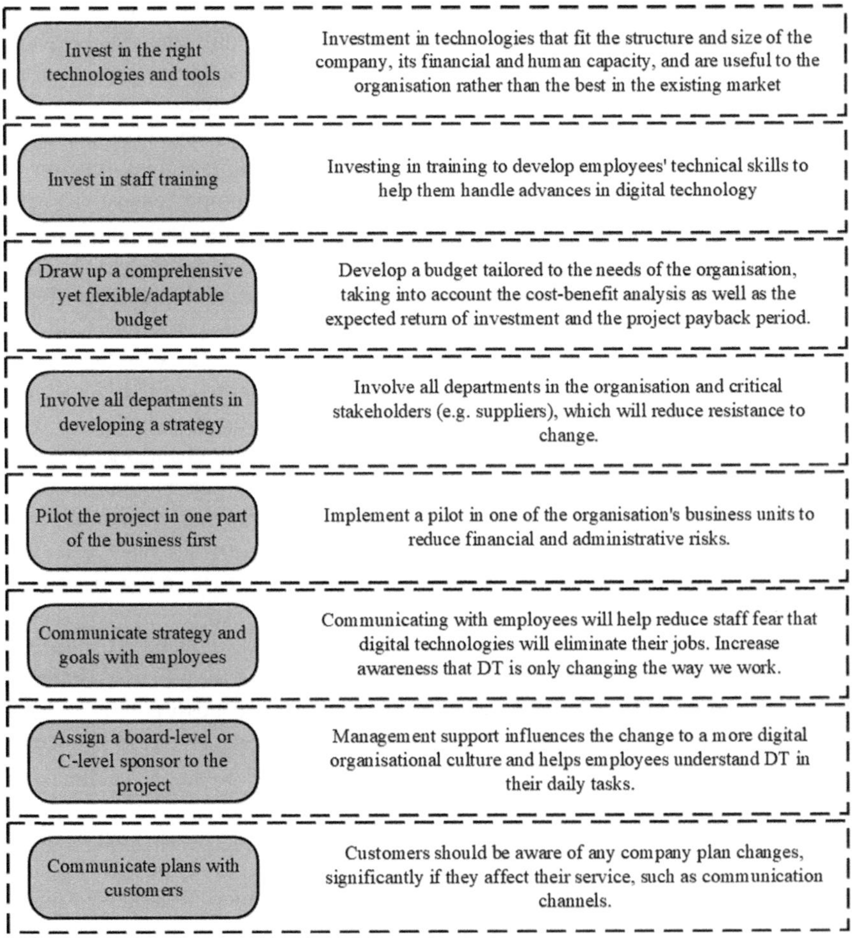

Fig. 3.15 Good practices for DT implementation (based on [88])

of the 2018–2022 literature review presented here identified the challenges of digital transformation in supply chains serving agri-food, textiles, automotive, timber, steel, and public services, as well as in humanitarian supply chains.

3.2.2 *Assessing the Impact of Digital Transformation on the Logistics Processes of Supply Chain Participants*

Digital transformation critically affects participants' and entire supply chains' current and future operations. More and more companies in supply chains are investing in digital technologies to better respond to customer's changing needs and, above all, to the growing demand for tailored products. The processes carried out by supply chain participants need to keep up with emerging new service trends, both in logistics and sales and at the same time, they need to be increasingly integrated. Therefore, Stank et al. [99] found that the essential elements of supply chain digitisation are "thinking" (improved analytics), "seeing" (enhanced visibility), and "acting" (heightened operational flexibility and reduced cycle time).

The importance of digital transformation in supply chain management is confirmed by many studies reported in the literature focus on assessing the impact of implementing digital solutions on company processes. This impact is assessed primarily through the prism of the benefits and the risks associated with implementing digital technologies to support current processes. Studies describing the impact of transformation most often refer to supply chains serving a specific sector (sectoral evaluation) or a group of companies operating in a specific familiar environment, e.g. in a particular region (environmental evaluation).

An example of a sectoral assessment is a study by Khan et al. [100], in which the authors assessed the impact of digitalisation on fast, fair, and safe humanitarian supply chains. The results indicated that digitisation significantly predicts fast, fair, and secure humanitarian logistics systems. The lack of well-developed information-sharing systems increases the risk of fraud and damages public trust in charities. The implementation of digital technologies in humanitarian supply chains has a positive impact on the accountability, security, trust, transparency and timeliness of the activities carried out by its members. However, the authors point out that a significant risk is the lack of experience and skills of humanitarian workers in handling modern digital technologies.

An example of an environmental assessment can be found in the research conducted by Lee et al. [101], in which the authors assessed the impact of technology development and digital transformation on supply chains and organisational performance in Malaysian manufacturing companies. The results of this research confirmed that digitisation positively impacts business performance, as Malaysian manufacturers are aware. Implementing digital technologies can help local companies grow their business efficiently, improve the quality of their services, achieve competitive values in the market, stay ahead of changes in the sector served and reduce unnecessary expenses. The key findings of the research clearly indicate that implementing digital transformation at the supply chain level improves the efficiency of the entire chain. However, more broadly, it also helps its participants to achieve high business performance and sustained growth. However, Lee et al. note that most manufacturers in Malaysia still need to gain knowledge and awareness of the benefits of implementing digital technologies in their operations.

Implementing digital technologies is intensely focused on improving communication within organisations and between supply chain partners. The implemented solutions enable companies to overcome the constraints of remote locations and the difficulty of exchanging data in real-time. For this reason, some of the studies assessing the impact of digital transformation focus on aspects concerning the impact of digital technologies on logistics integration, especially in terms of global supply chains. Indeed, many companies see digitalisation as enhancing their international capabilities [102]. In particular, digital transformation can benefit SMEs by improving their ability to compete with larger organisations through international collaboration opportunities [103]. An example of such publications is the study by Dethine et al. [104], who assessed the impact of digitalisation on the internationalisation capacity of SME companies in their analysis. Their results showed that the digitisation of SME enterprises allows them to reduce the costs of entering international markets by reducing the costs associated with spatial distance and export intermediaries. In addition, these enterprises optimise their business models by implementing digital solutions and improving close and direct connections with customers. However, the digitalisation of these companies also generates certain risks, such as potential damage to technical systems and data files, financial losses, loss of productivity and loss of confidential customer information.

A critical area of research related to the impact of digital transformation on the operations of companies and supply chains is to assess the impact of digital technologies on building resilience across the supply chain. Some authors point out that the impact of digitalisation on the resilience of logistics processes and the entire supply chain is a complex issue [105]. The research presented in [106] found that the development of digitalisation in individual links in the chain positively affects their level of resilience. This research also confirms that achieving successive levels of digital maturity positively impacts the resilience of a given chain. However, the strength of this impact already depends on the size of the company's turnover. The divergence of this impact between companies with the highest turnover versus those with low turnover is significant.

Using data analytics to support operational and predictive activities is critical to building resilience. According to Ivanov and Dalgui [25], data analytics should achieve the required level of visibility, improve forecast accuracy, reduce disruption to information flows, and improve contingency plans. In addition, using advanced trace and tracking systems, risks associated with missed delivery dates can be reduced, and contingency plans can be activated in real time (short response time to disruption). Blockchain digitisation and additive manufacturing can minimise the need for supply risk-limiting inventories, the reservation of production/operational capacity, and the associated costs of reserve resources. At the same time, the combination of big data with advanced trace&tracking systems and blockchain enables a proactive process of tracking the sources of disruptions, then observing the propagation of disruptions to ultimately build the resilience of the organisation at different planning levels [25]:

- development of short-term stabilisation measures through a complete understanding of available resources and inventories (creation of contingency plans);
- development of repair and improvement policies with a medium-term impact horizon;
- analysing and considering the long-term impact of incident-related effects on company performance.

The conclusions presented by Ivanov and Dalgui also support the findings of Pettite et al. [107], who demonstrate that technologies such as cloud computing and blockchain improve visibility, adaptability and anticipation, which translates into building supply chain resilience. The positive impact of selected I4.0 solutions on building supply chain resilience is shown in Fig. 3.16. Zouari et al. assessed this impact based on their literature review.

Significant factors in the impact of digital transformation on logistics processes and supply chain performance are the potential benefits that companies can realise from adopting digital technologies and changing business models. A literature review conducted by Büyüközkan and Göçer [86] allowed the authors to distinguish 15 benefits that determine the success of implementing digital transformation at the level of supply chain operations. These benefits are presented in Table 3.7.

3.2.3 Challenges of Implementing Digital Transformation Among Supply Chain Participants

Digital transformation concerns a critical change from the point of view of top management, middle managers, operational staff and external stakeholders working closely with the company. It involves different levels of management and different spheres of activity. For this reason, its implementation at the company and supply chain level involves considering specific challenges in the planned activities, which may determine the success or failure of the implemented projects and changes. A literature review in this area identified the most commonly described challenges related to implementing digital transformation in organisations.

In their study, Xu and Choi [108] identified the main challenge regarding implementing digital supply chain concepts. This is the requirement to collect data from multiple sources, to ensure that the information collected is up-to-date and reliable, and to develop a software architecture and platform that supports using this data to manage operations along the supply chain. At the same time, it is crucial to consider that supply chains comprise many internal and external entities. Therefore, the flow of information in this case can be prolonged and more error-prone

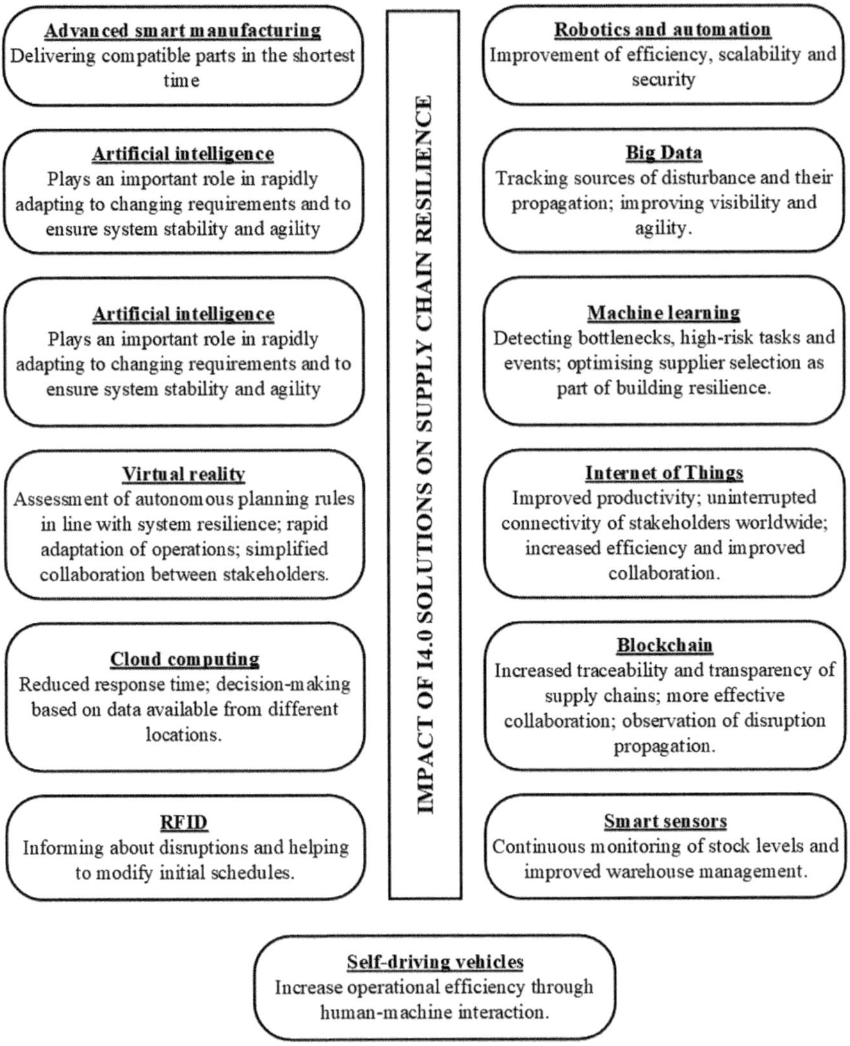

Fig. 3.16 The impact of implementing I4.0 solutions on the resilience of supply chains (based on [106])

than in the internal information flow of a single organisation. Research on the challenges of implementing digitisation and Industry 4.0 tools was also investigated in [109]. The results of his research indicated that collaboration and coordination are the most significant challenges to adopting Industry 4.0 out of the fourteen identified challenges, followed by resistance to change and governmental support. A significant challenge identified in the digital transformation process is the selection of appropriate digital tools and technologies to meet the needs of modern supply chains. Research in this area has been conducted by Ahi et al. [110], among others,

Table 3.7 Benefits of successful implementation of digital transformation at the supply chain level (based on [86])

Benefit	Description
Real-time visibility	Improving material flow management by providing interactive, secure and continuous visibility along the supply chain
Continuous collaboration	Ensuring better coordination of processes and continuity of collaboration between the supply chain actors involved
Alignment of suppliers	Integrating the interests of all supply chain participants with its objectives to create incentives for higher performance and build trust
Integration	Integration of digital information to provide supply chain partners with comprehensive information on inventory and material flows along the chain
Shared information	The ability to share sales forecast information and production data more quickly and efficiently
Highly evolved operating models	Quickly modify product and service functionality to flexibly respond to changing customer needs
Adopting advanced analytics and analytics tools	The ability to improve decision-making processes by providing information from advanced data analyses, as well as a better understanding of phenomena and solving previously unknown problems
Automated execution	Increasing operational efficiency through human–machine collaboration
Enhanced and accelerated innovation	Digital transformation inspires and supports innovation in projects, operations and customer relationship development
Maximum efficiency	It is achieved through full integration of people, processes and technology
Organisational flexibility	Digital plug-and-play capabilities make it easier to configure and reconfigure
Personalised experiences, customer-centric	The delivery network focuses on delivering personalised products and services
Enhanced responsiveness	The ability to respond faster to competitive actions, technological changes and signals of changing market demand

(continued)

Table 3.7 (continued)

Benefit	Description
Proactive prevention	Decision-making processes based on the results of predictive analyses improve the adaptability and reliability of systems
Last mile postponement	Effective management of company assets enables flexible adaptation of supply to changing demand

who analysed the extent of implementation of new technologies and the associated challenges vis-à-vis the operation of international supply chains.

Büyüközkan and Göçer also identified the challenges related to the digital transformation process. The literature research conducted by these authors allowed to identify the challenges that appear most frequently in publications on implementing digital technologies. These include [86]:

- Lack or limited preparation of plans in companies, lack of guidelines and tools to support planning processes.
- Lack of or limited cooperation with supply chain partners and integration of internal company functions.
- Lack of information exchange and reluctance of supply chain participants to share information.
- An organisation's pursuit of a "silver bullet"—basing decisions on the assumption that everything will be fine.
- Lack of knowledge and training to develop skills in new supply chain management forms.
- Limited agility and flexibility; lack of the required level of supply chain flexibility and agility.
- Lack of knowledge and skills to deal with the volatility of the environment and the supply chain's response to that volatility.
- Over-reliance on suppliers and lack of need to monitor material flows and terms and conditions of collaboration.
- Limited understanding of integrating digital and physical (non-digital) solutions to support supply chain management.

It is worth noting that the challenges of implementing digital transformation may vary depending on the industry the chain serves [83]. An example of research focused on the challenges of digital transformation in a specific sector is presented in [111]. In their analyses, the authors attempted to identify the challenges of implementing digital technologies in the agri-food sector. Implementing digitalisation is a challenge primarily for companies in the SME sector. For this reason, [Omrani], in his research, identified factors that determine the implementation of digital transformation, which are specific to SMEs. At the same time, however, Heavin and Power [112] emphasise that these challenges apply not only to SME businesses but also to large market players. Obviously, the extent of the challenges present in these two groups of companies differs.

A literature review by Abdallah et al. identified four main categories of challenges related to the digital transformation process. These categories are shown in Fig. 3.17 and discussed below [85]:

(1) People

The success of implementing digital transformation is determined by the ability of employees and managers to use digital technologies to improve processes and increase the efficiency of implemented activities. Therefore, acquiring and developing digital competencies is becoming a necessity in modern organisations, significantly as the demand for such skills will increase in the years to come [113]. At the same time, some employers do not understand the need for employees to acquire new skills, including digital skills that allow them to go through the organisation's digital change process comfortably.

A critical element related to implementing new technologies is people's fear of losing their jobs to robotisation and automation of logistics operations. Many people

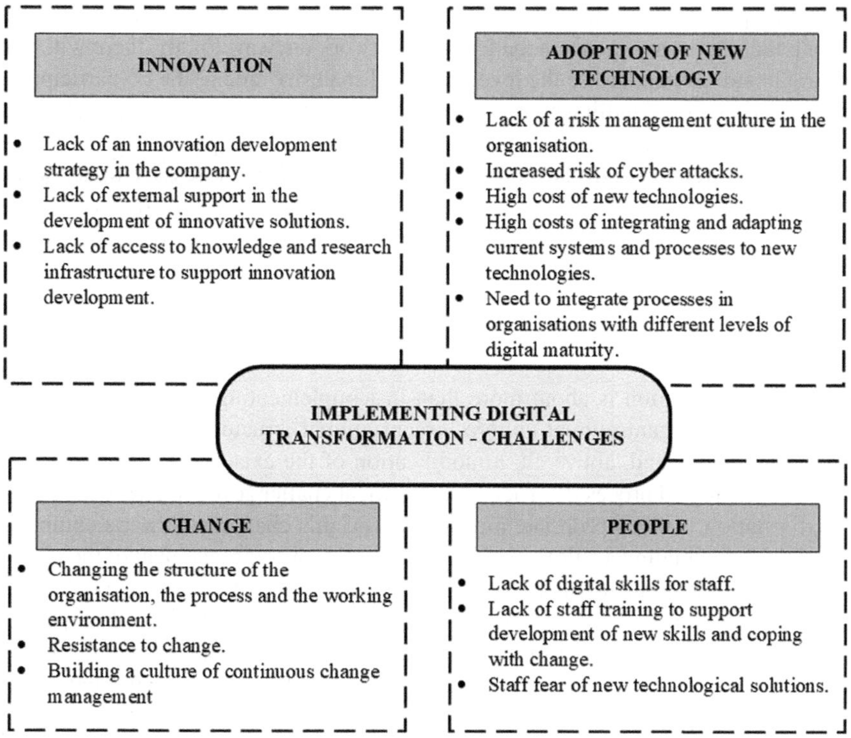

Fig. 3.17 Categories of challenges associated with the digital transformation process (based on [85])

also fear that they will be unable to work in complex cyber-physical systems, which will isolate them and result in their dismissal.

(2) Adoption of new technology

The implementation of new technology carries investment (financial) and organisational risks. Therefore, the adoption of new technology requires the introduction of a risk management culture in the organisation. This is particularly important as digital transformation increases the risk of cyber attacks. Therefore, building risk management systems and procedures to safeguard the company against these cyber threats is important.

Implementing new technologies means high investment costs for their purchase and implementation. At the same time, however, this implementation means that processes, procedures and implementation standards need to be adapted to the new working environment. At the same time, new technologies replace legacy solutions, which means migrating data from previous systems while adapting them to the requirements of new technologies. This is associated with the need for additional, high financial outlays, often not mentioned at the planning stage of the transformation processes.

It is also worth noting that implementing new technologies in individual links of the supply chain requires the adaptation of the effects of this implementation among the other participants in the logistics network anyway. Firstly, there will be a natural desire to standardise the level of digital maturity among the co-participants in the material flow handling process due to the desire to obtain the whole package of benefits of digital transformation. Secondly, even if some of the partners do not follow the accepted trend of introduced digital changes, it will be necessary to adapt the service standards to the requirements of exchange participants at different levels of digital maturity. Conversely, if the links that do not take up the digital transformation challenge are eliminated, there will be costs associated with changing the participants in the logistics network.

(3) Change

Digital transformation is about more than just implementing new IT solutions. It is a change of organisational culture, organisational structures, environment and working methods and, above all, a modification of the existing role of employees in the implemented processes. Therefore, a critical challenge associated with digital transformation is the appropriate management of this change within the company, particularly in building a culture based on the continuous improvement of the organisation and the accompanying need for continuous change. As has been emphasised many times in this book, digital transformation is an ongoing process that will not have an end as long as we observe technological developments in the enterprise's business environment. Therefore, it forces managers to manage their subordinate teams in such a way that, for their members, change is a natural part of their working conditions. If this approach is not taken, employees who have been taught to work according to the rules of the old system will strongly resist change, which forces them to step out of their comfort zone and take on new challenges. This can result in a prolonged change process and the need for additional costs and effort to achieve the desired goals.

(4) Innovation

Innovation is an essential success factor for successfully implementing digital transformation in enterprises. Therefore, an organisation's digitisation strategy should be integrated with the company's innovation development strategy. The new digital technologies implemented in the organisation should support and impulse the development of innovative solutions for implementing processes (process innovation) and creating new products/services (product innovation) that meet customers' changing needs. It is also worth noting that external funding bodies can support the implementation of innovation in companies. This makes it possible to reduce the organisation's own financial involvement in digital transformation processes, which can benefit from various types of external subsidies (e.g., European Union funds). This fact is significant from the point of view of companies belonging to the SME group, for which limited own financial resources are the main barrier to implementing digital transformation in their operations [87]. Another critical challenge in this area is access to knowledge and research infrastructure, which supports developing new initiatives. For this reason, the industry must cooperate with research units, which, with their activities, can support the implemented digital transformation processes in companies by developing new, personalised methods and tools for using digital technologies in business operations.

3.3 Logistics 4.0 Concept

The changes brought about by the development of Industry 4.0 tools are dynamic in nature and involve not only investment in new technologies but also radical changes to process models and the business environment. This means that the scope of impact of the I4.0 concept and the area of implementation of new technologies are no longer limited to production processes alone; they also apply to many supporting processes. The support area that ranks second after production processes in terms of the number of I4.0 solutions implemented is logistics. The critical importance of digital transformation and I4.0 solutions in contemporary logistics systems is confirmed by the fact that the phrase Logistics 4.0 has entered the vocabulary of terms describing contemporary changes in industrial processes.

At present, relatively few publications in the literature refer to the concept of Logistics 4.0. However, the literature review presented by Tubis and Grzybowska in 2023 [98] and its comparison with the review prepared by Winkelhaus and Grosse in 2020 [114] indicates that the concept is steadily developing and growing in popularity not only in industry but also in research. This is also confirmed by the number of publications referring to these solutions, which has steadily increased since 2017 [115]. Of course, this growth is due to the continuous development of the Industry 4.0 concept and the increasing popularity of applied I4.0 solutions, which find application not only in the area of production but, precisely above all, in logistics processes. Today, Logistics 4.0 is recognised as one of the fundamental pillars of Industry 4.0 because of its strong integration with production processes.

Logistics 4.0 continues to be defined in different ways, depending on whether it refers to processes, systems or its role in the functioning of supply chains. An interesting proposal for defining the term Logistics 4.0 is presented in [116], where the author defined it as a collective term for technologies and concepts of value chain organisation, in which cyber-physical systems monitor the processes of physical material flow, create a virtual copy of the physical world and make decentralised decisions. The universal definition of the concept can be considered the one formulated by Winkelhaus and Grosse, who described Logistics 4.0 as a logistics system that makes it possible to sustainably meet the personalised needs of internal and external customers through digitally supported development [114]. Logistics 4.0 uses new technologies based on cyber-physical systems to support logistics processes [117]. This support is made possible by increased real-time data availability and simple and repetitive operations automation.

Szymanska et al. [118] identified two dimensions of the Logistics 4.0 concept in their research: (a) processual—internal and external processes are a subject of the Logistics 4.0 actions, and (b) technical—technologies and tools that support internal processes in the supply chains. Winkelhaus and Grosse, on the other hand, in their research distinguished three dimensions that make up the construction of the Logistics 4.0 concept [114]:

- The external dimension is described by paradigm shifts driven by increasing customer expectations for highly individualised products and the development of Industry 4.0, which supports globalisation and sustainable development processes. This dimension also encompasses social changes related to modifying consumption behaviour and customers' expected access to different types of information.
- The technological dimension consists mainly of the technological factors of Industry 4.0 but also aspects concerning digital transformation in the internal processes of organisations and entire supply chains. Technological elements enable the paradigmatic change that leads to a shift from traditional logistics systems to Logistics 4.0.
- Tasks, domains, and people characterise logistics processes' dimension. Tasks refer to management activities and execution activities. These activities are assigned to four domains according to the direction of the material flow implemented, which refer to procurement logistics, internal (production) logistics, distribution logistics and reverse logistics. The human factor refers to employee behaviour and managerial decisions that influence the risk, execution quality and efficiency of logistics processes. At the same time, it should be borne in mind that the I4.0 solutions being introduced are not intended to replace man but to support his activities within the framework of man–machine cooperation.

Based on their research, the authors also concluded that the main driver of the evolution of Logistics 4.0 is external change, and its causes can be divided into three categories: customer-driven, competitor-driven and environment-driven [114].

Information and access are central to all the changes associated with the fourth industrial revolution concerning logistics processes. For this reason, the technologies being implemented can be grouped into three categories [114]:

- Technologies for information generation—provide the visibility required to improve planning and control of temporally and spatially distributed, complex systems. Previously, physical and information flows were often handled separately, leading to inefficiencies. Using solutions such as IoT in conjunction with RFID enables the required visibility, which can be used to improve most logistics operations directly or indirectly.
- Technologies for information processing. Processing the large amount of data generated is a complex process necessary for the information collected to support managers in their decision-making processes. A solution that can meet this challenge is Big Data analytics, which focuses on processing massive data and drawing conclusions from the analyses performed. Blockchain technology, which can support the transaction handling process, can also fall into this category.
- Information-enabling technologies. The information collected should be actively used to improve logistics processes and manage material flows. In this regard, two possibilities should be considered: (a) the system can use the information to execute logistics operations, e.g. by cyber-physical systems, or (b) in complex planning tasks, humans can use the information in decision-making processes, e.g. Big Data to optimise inventory levels or support employees in executing operations using augmented reality.

Logistics 4.0 uses technological tools whose development is supported by Industry 4.0, which are characterised in this chapter in Sect. 3.1. However, this list should be supplemented by several solutions dedicated to supporting logistics processes. Figure 3.18 shows the solutions most commonly used in Logistics 4.0 systems.

The primary logistics system supported by Industry 4.0 solutions is the warehouse. It is the warehouse handling processes that were first supported by new technologies such as automatic data identification (bar codes, radio frequency identification—RFID, QR codes) and autonomous vehicles (autonomous mobile robots, unmanned aerial vehicles—drones) [119].

3.3.1 Radio Frequency Identification (RFID)

RFID is a technology dedicated to tracking and tracing goods, assets and accompanying information flows. RFID encompasses all technological solutions automatically using radio waves to identify objects or people [120]. RFID technologies respond to the need for intelligent identification of various objects and are therefore

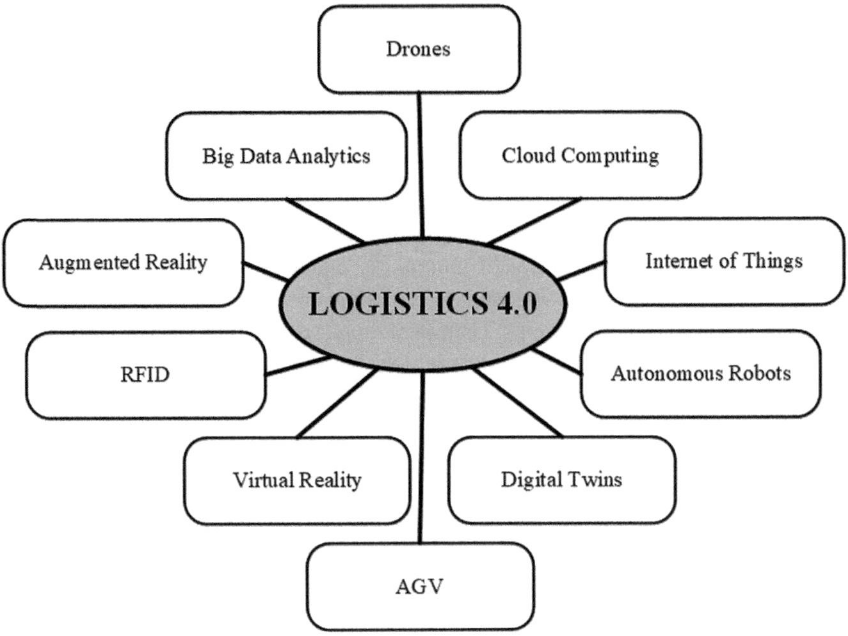

Fig. 3.18 Industry 4.0 solutions applied to logistics processes

used in many industries, such as retail [121], logistics [122], building and construction [123], fashion [124], and healthcare [125].

Several methods of RFID object identification can be found in literature and practice. Most available solutions consist of a reader and a tag, which include an antenna and a microchip. The role of the microchip is to store information about the object, while the antenna enables the microchip to transmit information to the reader. There are two types of tags [126]:

- Passive—the most straightforward and cheapest information carriers that do not have an embedded information source.
- Active—have their own power supply, thanks to which they have internal read and write capabilities and can transmit signals over long distances.

The industry also has semi-passive tags and intermediate solutions to the two primary technologies. It is also worth noting that RFID technology can support both continuous and discrete object tracking [126].

The increasing popularity of RFID technology in logistics is determined by the many benefits associated with its implementation. Chanchaichuji et al. [127] identified six main benefits of using RFID in logistics based on a literature review. These benefits are shown in Fig. 3.19.

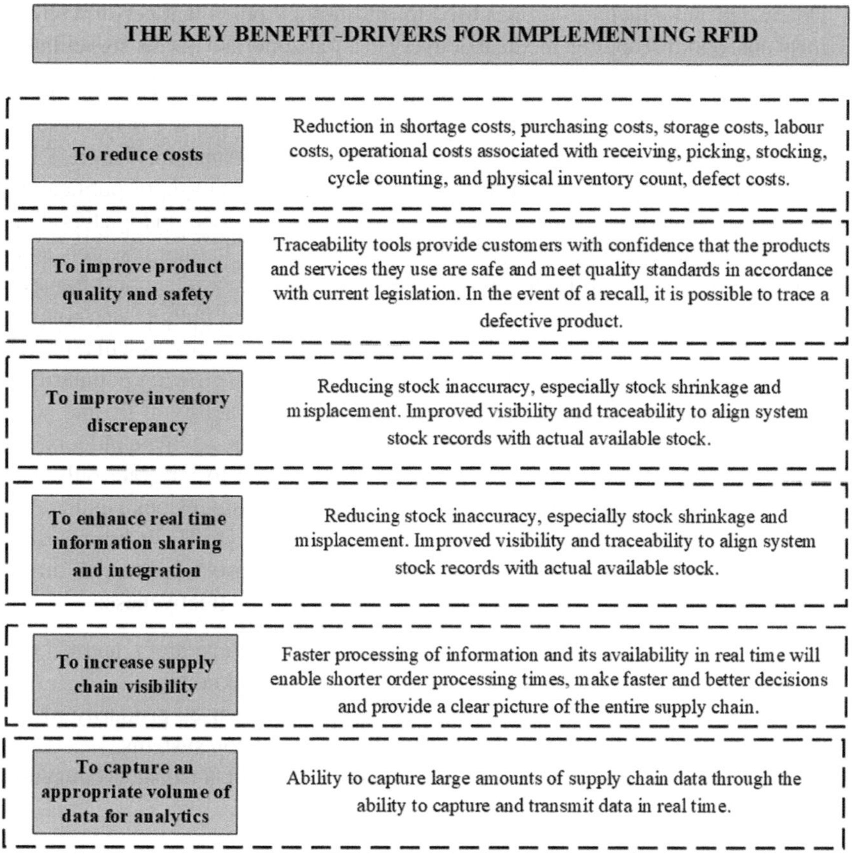

Fig. 3.19 Benefits of implementing RFID technology (based on [127])

3.3.2 Unmanned Aerial Vehicles (UAV)—Drones

Unmanned aerial vehicles, also known as drones, were initially conceived as a technology used for various military missions [128]. However, Industry 4.0 has also resulted in their use in non-military areas. Their area of application is one of many criteria for their classification. Singhal et al. [129] divided drones into three groups about the missions they perform:

- Civilian group—includes drones used primarily for industrial and social purposes, e.g., mining, agriculture, construction, maintenance, logistics, delivery or disaster management.
- Environment group—includes drones used to monitor ecosystems such as soil, crops, water, underwater, and mountains or to measure air quality, pollution, and gas concentrations.

- Defence group—includes drones used for military purposes that perform spying missions, bomb dropping, medical delivery in a war zone, and border surveillance.

Using drones as an enabling technology for logistics operations is relevant to the book's topic. Yoo et al. [130] point out that the popularity of drone use in this case has been mainly influenced by their use by large online retailers (Amazon, Google, DHL, and Walmart) in the parcel delivery process. However, an analysis of publications related to the use of drones indicates that the primary use of drones in logistics at this point is mainly in support of selected warehouse operations. In their research, some authors even indicate that drones will be a crucial technology to support transport processes in the internal logistics of smart factories [131]. Many authors also note that the widespread implementation of UAVs to support logistics processes will represent future trends in developing Logistics 4.0 systems [132]. The growing popularity of drones in logistics processes is mainly determined by their ability to fly and hover autonomously, avoid obstacles in different warehouse layouts, navigate indoors, and land precisely [119]. In particular, their lack of restriction to navigate only on a flat surface and their applicability in all three warehouse dimensions significantly influences the search for their increasing use in logistics [132]. In his research, Wawrla distinguished three main warehouse processes whose support with drones can bring significant increases in operational efficiency and effectiveness:

- Inventory management—drones can be used here for inventory audit, cycle counting, item search, buffer stock maintenance, and stocktaking.
- Intralogistics of items—in this process, drones should be used primarily for the express delivery of tools and spare parts feeding production systems.
- Inspection and surveillance—drones can be used to monitor and inspect dangerous areas or high altitudes or counter undesirable behaviour.

The literature review and implementation reports presented in [133] indicate numerous benefits and challenges associated with using drones in warehouse operations. The most commonly identified benefits and challenges are shown in Fig. 3.20.

The critical constraint at present is primarily legislative challenges. These are the ones that significantly limit the applicability of drones in external deliveries in many countries. For this reason, their full potential must be realised in internal logistics. In Poland, several projects are currently underway to research the use of drones in warehouse logistics and support monitoring systems for transport and logistics infrastructure. The results obtained from these studies (e.g. [133, 135, 136]) confirm the excellent potential for improving material handling processes through the use of AUV systems.

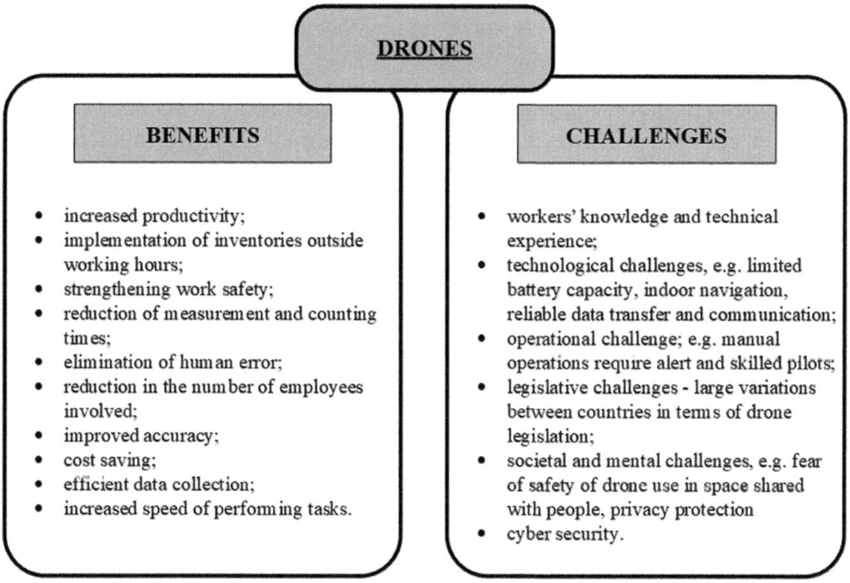

Fig. 3.20 Benefits and challenges of drone use in warehouse operations (based on [133, 134])

3.3.3 Automated Guide Vehicle (AGV)

Drones are not the only handling devices that owe their intensive development to the fourth industrial revolution. Far more popular in internal logistics (warehousing and manufacturing) are AGV systems, which can refer to automated or autonomous guide vehicles. These systems are used in production plants, warehouses, distribution centres and transhipment terminals. Automated guide vehicles have been known for a long time. However, with the development of the Logistics 4.0 concept, companies have increased their interest in implementing large-scale internal material flow handling processes. The growing interest in implementing AGV systems in logistics processes is also influenced by the numerous benefits described in the literature and industry reports. Among the most important are [137, 138]:

- reduced labour costs,
- increased safety,
- increased operational accuracy,
- increased productivity,
- elimination of human errors,
- sustainability of logistics operations supporting material flows in supply chains,
- reduction of energy consumption, particularly for electrically powered AGVs,
- reduction of atmospheric emissions of Particle Matter and Greenhouse Gasses like CO_2 and NO_2,

- improvement of human safety above all by reducing the number of accidents in warehouse processes involving workers.

Implementing an AGV system in internal logistics is always associated with high investment costs [139]. However, the investment incurred for implementation usually pays off very quickly due to the savings generated by AGVs during their operational phase. The most significant savings associated with the use of AGV systems can be achieved through [140]:

- the ability to operate 24/7 with minimal labour costs and human intervention,
- reduction in overtime pay,
- lower maintenance expenses compared to conventional vehicles,
- increased productivity in logistics operations and extended service levels for the entire SC,
- indirect cost savings resulting from improved safety and accident reduction.

The increasing use of AGV systems in logistics operations has increased the variety of tasks these vehicles perform, and their scope is no longer solely concerned with transport. The findings of DeRyck et al. [141] indicate that fleets of AGVs being designed will need to be more heterogeneous in order to be able to adapt dynamically and flexibly to the tasks assigned to them. Each vehicle will be assigned a specific functionality to operate in a flexible plug-and-produce system. AGVs can be used in various operational situations by equipping the vehicles to perform specific tasks. However, this is not the only direction of development for these systems. DeRyck et al. emphasise in their research that the future development of AGVs should take into account the following trends [141]:

- the growing demand for customised AGVs and expansion of their functionality;
- the development of flexible production systems;
- the adaptation of AGV systems to the specific needs of SME companies;
- the building of increasingly large and complex systems that should meet new needs for transport in the factory.

AGV systems are primarily used in handling, picking, packing, and palletising operations [142]. Their operation's design and control processes are complex, as they require considering many elements related to the operation of vehicles in coordinated material handling processes carried out in a space shared with people. Therefore, important aspects related to the planning and control of their operations include [143]:

- guide-path design,
- idle vehicle positioning,
- vehicle scheduling,
- battery management,
- vehicle routing and deadlock resolution,
- estimating the number of vehicles required (or determining vehicle requirements).

Optimisation of the plans developed for the above aspects of AGV management is the subject of numerous studies reported in the literature. This is mainly due to the desire to exploit the full potential of replacing traditional handling equipment with automated solutions. However, it is worth noting that using AGVs in anthropotechnical systems generates new hazards and changes the risk levels associated with cargo handling. Some of the adverse events identified are due to errors in the design of the system operation and planning of its activities. However, a set of undesirable events already occur during the system operation phase and cannot always be counteracted. Therefore, research directed towards developing guidelines for the operation and maintenance of AGV systems that will support logistics processes that are resilient to disruptions and implemented in a safe working environment is necessary. The safety and operational efficiency of AGV systems will determine their further development and broader application in the logistics handling processes of cargo in supply chains.

3.4 Summary

As already indicated in Chap. 1, an important phenomenon influencing the creation of a new risk paradigm is the digital transformation and the related fourth industrial revolution. Industry 4.0 represents a de facto revolution, as the solutions it introduces have entirely changed the working environment, the way processes are carried out, and even the business models implemented by individual organisations, as well as entire supply chains in all sectors of the economy. The ability to use large databases from various sources and available in many cases in real-time has created enormous potential for quantitative and qualitative analyses to support the decision-making processes of managers and management. Thanks to the availability of a wealth of information on historical phenomena and events and the ability to study their correlation with company performance, managers have gained knowledge that improves their ability to identify adverse events and assess the risk of their occurrence. Thanks to the continuous transfer of real-time data, they can react faster and better to unexpected disruptions and more accurately and efficiently adjust ways/tools to mitigate risks. However, the implementation of digital solutions and automation also brings with it the emergence of new risks that were not present in traditional systems, such as cyber-attacks, simulation sickness, and technological exclusion, for example. This makes the risk assessment scope necessary to consider a broader analytical context and go beyond the existing boundaries. The aim of this chapter was to present the tools of Industry 4.0 and the impact of their implementation and digital transformation on changes in logistics processes and supply chains.

Industry 4.0 has introduced new solutions into the business space. However, it has also benefited from previously known technologies, which have boosted intensive development by combining them into cyber-physical systems that operate in the new industrial reality. Therefore, to understand the need for a new approach to risk management, it is first necessary to understand these technological solutions and the principles of their application in Industry 4.0 systems. Therefore, Sect. 3.1 characterises the most important tools and technologies used in complex cyber-physical systems, outlining not only the assumptions for their application but also the benefits and risks involved. The principles of implementing Industry 4.0 solutions and their impact on modifying internal processes and business models implemented in companies were also presented. In particular, attention was paid to the change in man's role in processes supported by I4.0 technologies and the need to change employees' competencies. The acquisition by employees of so-called digital competencies, allowing them to function efficiently in a dynamically changing work environment, is one of the critical challenges related to the implementation but also to the use of the full potential provided by I4.0 tools.

Industry 4.0 has also accelerated digital transformation processes, which has a very significant impact on the execution of internal processes but, above all, on the relationships between participants in supply chains. This impact is so substantial that a new concept called Digital Supply Chain has been formulated based on this trend. Section 3.2 presents its basic assumptions and the conceptual framework for implementing digital transformation processes in supply chain links. Research in this area shows no one-size-fits-all approach to implementing digital transformation in supply chains. Therefore, it is necessary to build an implementation strategy that considers the specifics of the sector served, the requirements reported by consumers, and the supply chain structure itself, including the number of links at each level of the logistics network and the vertical and horizontal relationships linking them. The second important issue related to the digital transformation of modern supply chains is to assess its impact on changes in the processes and systems of actors involved in handling material flows. In their research, many authors focus primarily on the benefits associated with the digital transformation of organisations, looking for arguments supporting the validity of its implementation and continuous development. Notable benefits in favour of its implementation include, first and foremost, improving communication within the organisation and with business partners, as well as building resilient organisations and supply chains. However, digitalisation and automated logistics processes also generate new challenges and risks for companies. This is because it no longer only concerns the implementation of new information and communication technologies in operational activities but also modifies the organisational culture, methods of carrying out assigned tasks, and even business models implemented at the level of strategic changes in the organisation. A critical challenge many enterprises face is the need for more organisational preparation for the changes enforced by the modern market. The most critical category in this case is people with very limited digital skills and resistance to change. Failure to prepare them properly additionally results in fear of losing their jobs and blocking the implementation of new solutions. No less importantly, the very process of adapting new technology in an organisation is also a challenge. In this research, there are analyses of the high costs of implementation, which are related not only to the purchase of new technology but also to its integration with previous systems or the migration of data to new systems. Adapting new technology also requires changes to processes and operating procedures, particularly developing new ways of dealing with emerging risks, such as cyber-attacks. All these activities force intensive and dynamic changes in the company, which requires building an appropriate organisational culture based on flexible adaptation to new operating conditions and extensive use of the risk management potential associated with these changes.

The development of Industry 4.0 in recent years has meant that processes relating to digitisation and automation have moved beyond the boundaries of production processes and are now also being implemented with comparable success in other processes, including activities related to the logistical handling of material flows. For this reason, Sect. 3.3 presents the concept of Logistics 4.0, now considered one of the fundamental pillars of Industry 4.0. Indeed, logistics processes are the second largest area of application of digital technologies and automation at the level of both material flow handling and information handling. Almost all of the solutions used in intelligent manufacturing are also applied to logistics processes in individual companies and their cooperation in the supply chain. These technologies are also complemented by automatic and autonomous transport solutions, which provide essential support for cargo movement processes, nowadays primarily within internal logistics. These solutions will also be the subject of further research presented in the following chapters of this book.

References

1. Oztemel E, Gursev S (2020) Literature review of Industry 4.0 and related technologies. J Intell Manuf 31:127–182. https://doi.org/10.1007/s10845-018-1433-8
2. Roblek V, Meško M, Krapež A (2016) A complex view of Industry 4.0. Sage Open 6, 215824401665398. https://doi.org/10.1177/2158244016653987
3. Cañas H, Mula J, Díaz-Madroñero M, Campuzano-Bolarín F (2021) Implementing Industry 4.0 principles. Comput Ind Eng 158:107379. https://doi.org/10.1016/j.cie.2021.107379
4. Ortt R, Stolwijk C, Punter M (2020) Implementing Industry 4.0: assessing the current state. J Manuf Technol Manage 31:825–836. https://doi.org/10.1108/JMTM-07-2020-0284
5. Oesterreich TD, Teuteberg F (2016) Understanding the implications of digitisation and automation in the context of Industry 4.0: a triangulation approach and elements of a research agenda for the construction industry. Comput Ind 83:121–139. https://doi.org/10.1016/j.compind.2016.09.006
6. Stock T, Seliger G (2016) Opportunities of sustainable manufacturing in Industry 4.0. Procedia CIRP 40:536–541. https://doi.org/10.1016/j.procir.2016.01.129
7. Osterrieder P, Budde L, Friedli T (2020) The smart factory as a key construct of industry 4.0: a systematic literature review. Int J Prod Econ 221:107476. https://doi.org/10.1016/J.IJPE.2019.08.011
8. Roblek V, Meško M, Pušavec F, Likar B (2021) The role and meaning of the digital transformation as a disruptive innovation on small and medium manufacturing enterprises. Front Psychol 12. https://doi.org/10.3389/fpsyg.2021.592528
9. Posada J, Toro C, Barandiaran I, Oyarzun D, Stricker D, de Amicis R, Pinto EB, Eisert P, Dollner J, Vallarino I (2015) Visual computing as a key enabling technology for Industrie 4.0 and industrial internet. IEEE Comput Graph Appl 35:26–40. https://doi.org/10.1109/MCG.2015.45
10. Lu Y (2017) Industry 4.0: a survey on technologies, applications and open research issues. J Ind Inf Integr 6:1–10. https://doi.org/10.1016/j.jii.2017.04.005

11. Schroeder A, Trease B, Arsie A (2019) Balancing robot swarm cost and interference effects by varying robot quantity and size. Swarm Intell 13:1–19. https://doi.org/10.1007/s11721-018-0161-1
12. Nosalska K, Piątek ZM, Mazurek G, Rządca R (2019) Industry 4.0: coherent definition framework with technological and organizational interdependencies. J Manuf Technolo Manage 31:837–862. https://doi.org/10.1108/JMTM-08-2018-0238
13. Veile JW, Kiel D, Müller JM, Voigt K-I (2019) Lessons learned from Industry 4.0 implementation in the German manufacturing industry. J Manuf Technol Manage 31:977–997. https://doi.org/10.1108/JMTM-08-2018-0270
14. Moktadir MdA, Ali SM, Kusi-Sarpong S, Shaikh MdAA (2018) Assessing challenges for implementing Industry 4.0: implications for process safety and environmental protection. Process Saf Environ Protect 117:730–741. https://doi.org/10.1016/j.psep.2018.04.020
15. Qin J, Liu Y, Grosvenor R (2016) A categorical framework of manufacturing for Industry 4.0 and beyond. Procedia CIRP 52:173–178. https://doi.org/10.1016/J.PROCIR.2016.08.005
16. Hermann M, Pentek T, Otto B (2016) Design principles for Industrie 4.0 scenarios. In: 2016 49th Hawaii international conference on system sciences (HICSS). IEEE, pp 3928–3937. https://doi.org/10.1109/HICSS.2016.488
17. Ghobakhloo M, Ching NT (2019) Adoption of digital technologies of smart manufacturing in SMEs. J Ind Inf Integr 16:100107. https://doi.org/10.1016/j.jii.2019.100107
18. Bosman L, Hartman N, Sutherland J (2019) How manufacturing firm characteristics can influence decision making for investing in Industry 4.0 technologies. J Manuf Technol Manage 31:1117–1141. https://doi.org/10.1108/JMTM-09-2018-0283
19. Kiel D, Müller JM, Arnold C, Voigt K-I (2017) Sustainable industrial value creation: benefits and challenges of Industry 4.0. Int J Inno Manage 21:1740015. https://doi.org/10.1142/S1363919617400151
20. Hecklau F, Galeitzke M, Flachs S, Kohl H (2016) Holistic approach for human resource management in Industry 4.0. Procedia CIRP 54:1–6. https://doi.org/10.1016/j.procir.2016.05.102
21. Erol S, Jäger A, Hold P, Ott K, Sihn W (2016) Tangible Industry 4.0: a scenario-based approach to learning for the future of production. Procedia CIRP 54:13–18. https://doi.org/10.1016/j.procir.2016.03.162
22. Cusumano MA, Kahl SJ, Suarez FF (2015) Services, industry evolution, and the competitive strategies of product firms. Strateg Manag J 36:559–575. https://doi.org/10.1002/smj.2235
23. Jiang P, Ding K, Leng J (2016) Towards a cyber-physical-social-connected and service-oriented manufacturing paradigm: social manufacturing. Manuf Lett 7:15–21. https://doi.org/10.1016/J.MFGLET.2015.12.002
24. Pereira AC, Romero F (2017) A review of the meanings and the implications of the Industry 4.0 concept. Procedia Manuf 13:1206–1214. https://doi.org/10.1016/j.promfg.2017.09.032
25. Ivanov D, Dolgui A (2019) New disruption risk management perspectives in supply chains: digital twins, the ripple effect, and resileanness. IFAC-PapersOnLine 52:337–342. https://doi.org/10.1016/j.ifacol.2019.11.138
26. Kamble S, Gunasekaran A, Dhone NC (2020) Industry 4.0 and lean manufacturing practices for sustainable organisational performance in Indian manufacturing companies. Int J Prod Res. 58:1319–1337. https://doi.org/10.1080/00207543.2019.1630772
27. Bragança S, Costa E, Castellucci I, Arezes PM (2019) A brief overview of the use of collaborative robots in Industry 4.0: human role and safety. In: Arezes P, Baptista JS, Barroso MP, Carneiro P, Cordeiro P, Costa N, Melo RB, Miguel AS, Perestrelo G (eds) Occupational and environmental safety and health. Studies in systems, decision and control. Springer, pp 641–650. https://doi.org/10.1007/978-3-030-14730-3_68
28. Schou C, Andersen RS, Chrysostomou D, Bøgh S, Madsen O (2018) Skill-based instruction of collaborative robots in industrial settings. Robot Comput Integr Manuf 53:72–80. https://doi.org/10.1016/J.RCIM.2018.03.008
29. Lee J, Bagheri B, Kao H-A (2015) A cyber-physical systems architecture for Industry 4.0-based manufacturing systems. Manuf Lett 3:18–23. https://doi.org/10.1016/j.mfglet.2014.12.001

30. Monostori L, Kádár B, Bauernhansl T, Kondoh S, Kumara S, Reinhart G, Sauer O, Schuh G, Sihn W, Ueda K (2016) Cyber-physical systems in manufacturing. CIRP Ann 65:621–641. https://doi.org/10.1016/j.cirp.2016.06.005
31. Tortorella GL, Fettermann D (2018) Implementation of Industry 4.0 and lean production in Brazilian manufacturing companies. Int J Prod Res 56:2975–2987. https://doi.org/10.1080/00207543.2017.1391420
32. Zhang Y, Huang T, Bompard EF (2018) Big data analytics in smart grids: a review. Energy Inform 1:8. https://doi.org/10.1186/s42162-018-0007-5
33. Elgendy N, Elragal A (2014) Big Data analytics: a literature review paper. In: Perner P (ed) Advances in data mining. Applications and theoretical aspects. ICDM 2014. Springer, pp 214–227. https://doi.org/10.1007/978-3-319-08976-8_16
34. Moura J, Serrão C (2015) Security and privacy issues of Big Data. In: Zaman N, Seliaman ME, Hassan MF (eds) Handbook of research on trends and future directions in Big Data and web intelligence. Information Science Reference, pp 20–52. https://doi.org/10.4018/978-1-4666-8505-5.ch002
35. Birje MN, Challagidad PS, Goudar RH, Tapale MT (2017) Cloud computing review: concepts, technology, challenges and security. Int J Cloud Comput 6:32. https://doi.org/10.1504/IJCC.2017.083905
36. Yen C-T, Liu Y-C, Lin C-C, Kao C-C, Wang W-B, Hsu Y-R (2014) Advanced manufacturing solution to industry 4.0 trend through sensing network and Cloud Computing technologies. In: 2014 IEEE international conference on automation science and engineering (CASE). IEEE, pp 1150–1152. https://doi.org/10.1109/CoASE.2014.6899471
37. Rose K, Eldridge S, Chapin L. The Internet of Things: an overview. Understanding the issues and challenges of a more connected world
38. Serror M, Hack S, Henze M, Schuba M, Wehrle K (2021) Challenges and opportunities in securing the industrial Internet of Things. IEEE Trans Industr Inform 17:2985–2996. https://doi.org/10.1109/TII.2020.3023507
39. Trappey AJC, Trappey CV, Hareesh Govindarajan U, Chuang AC, Sun JJ (2017) A review of essential standards and patent landscapes for the Internet of Things: a key enabler for Industry 4.0. Adv Eng Inform 33:208–229. https://doi.org/10.1016/j.aei.2016.11.007
40. Vieira J, Poças Martins J, Marques de Almeida N, Patrício H, Gomes Morgado J (2006) Towards resilient and sustainable rail and road networks: a systematic literature review on Digital Twins. Sustainability 14:7060. https://doi.org/10.3390/su14127060
41. Kritzinger W, Karner M, Traar G, Henjes J, Sihn W (2018) Digital Twin in manufacturing: a categorical literature review and classification. IFAC-PapersOnLine 51:1016–1022. https://doi.org/10.1016/j.ifacol.2018.08.474
42. Errandonea I, Beltrán S, Arrizabalaga S (2020) Digital Twin for maintenance: a literature review. Comput Ind 123:103316. https://doi.org/10.1016/j.compind.2020.103316
43. Loganathan A, Ahmad NS (2023) A systematic review on recent advances in autonomous mobile robot navigation. Eng Sci Technol Int J 40:101343. https://doi.org/10.1016/j.jestch.2023.101343
44. Rubio F, Valero F, Llopis-Albert C (2019) A review of mobile robots: concepts, methods, theoretical framework, and applications. Int J Adv Robot Syst 16:172988141983959. https://doi.org/10.1177/1729881419839596
45. Wijayathunga L, Rassau A, Chai D (2023) Challenges and solutions for autonomous ground robot scene understanding and navigation in unstructured outdoor environments: a review. Appl Sci 13:9877. https://doi.org/10.3390/app13179877
46. Kanishka K, Acherjee B (2023) Revolutionizing manufacturing: a comprehensive overview of additive manufacturing processes, materials, developments, and challenges. J Manuf Process 107:574–619. https://doi.org/10.1016/j.jmapro.2023.10.024
47. Rasiya G, Shukla A, Saran K (2021) Additive manufacturing-a review. Mater Today Proc 47:6896–6901. https://doi.org/10.1016/j.matpr.2021.05.181
48. Wong KV, Hernandez A (2012) A review of additive manufacturing. ISRN Mech Eng 2012:1–10. https://doi.org/10.5402/2012/208760

49. Saleh Alghamdi S, John S, Roy Choudhury N, Dutta NK (2021) Additive manufacturing of polymer materials: progress, promise and challenges. Polymers (Basel) 13:753. https://doi.org/10.3390/polym13050753
50. Hasanov S, Alkunte S, Rajeshirke M, Gupta A, Huseynov O, Fidan I, Alifui-Segbaya F, Rennie A (2021) Review on additive manufacturing of multi-material parts: progress and challenges. J Manuf Mater Process 6:4. https://doi.org/10.3390/jmmp6010004
51. Masoni R, Ferrise F, Bordegoni M, Gattullo M, Uva AE, Fiorentino M, Carrabba E, Di Donato M (2017) Supporting remote maintenance in Industry 4.0 through augmented reality. Procedia Manuf 11:1296–1302. https://doi.org/10.1016/j.promfg.2017.07.257
52. Song Y, Koeck R, Luo S (2021) Review and analysis of augmented reality (AR) literature for digital fabrication in architecture. Autom Constr 128:103762. https://doi.org/10.1016/j.autcon.2021.103762
53. Van Krevelen DWF, Poelman R (2010) A Survey of augmented reality technologies, applications and limitations. Int J Virt Real 9:1–20. https://doi.org/10.20870/IJVR.2010.9.2.2767
54. Daponte P, De Vito L, Picariello F, Riccio M (2014) State of the art and future developments of the augmented reality for measurement applications. Measurement 57:53–70. https://doi.org/10.1016/j.measurement.2014.07.009
55. Azuma R, Baillot Y, Behringer R, Feiner S, Julier S, MacIntyre B (2001) Recent advances in augmented reality. IEEE Comput Graph Appl 21:34–47. https://doi.org/10.1109/38.963459
56. Masood T, Egger J (2019) Augmented reality in support of Industry 4.0—implementation challenges and success factors. Robot Comput Integr Manuf 58:181–195. https://doi.org/10.1016/J.RCIM.2019.02.003
57. de Souza Cardoso LF, Mariano FCMQ, Zorzal ER (2020) A survey of industrial augmented reality. Comput Ind Eng 139:106159. https://doi.org/10.1016/j.cie.2019.106159
58. Syberfeldt A, Holm M, Danielsson O, Wang L, Brewster RL (2016) Support systems on the industrial shop-floors of the future—operators' perspective on augmented reality. Procedia CIRP 44:108–113. https://doi.org/10.1016/j.procir.2016.02.017
59. Danielsson O, Holm M, Syberfeldt A (2020) Augmented reality smart glasses in industrial assembly: current status and future challenges. J Ind Inf Integr 20:100175. https://doi.org/10.1016/j.jii.2020.100175
60. Hofmann B, Haustein D, Landeweerd L (2017) Smart-glasses: exposing and elucidating the ethical issues. Sci Eng Ethics 23:701–721. https://doi.org/10.1007/s11948-016-9792-z
61. Zigart T, Schlund S (2020) Evaluation of augmented reality technologies in manufacturing—a literature review. In: Nunes I (ed) Advances in human factors and systems interaction. AHFE 2020. Advances in intelligent systems and computing. Springer, pp 75–82. https://doi.org/10.1007/978-3-030-51369-6_11
62. Palmarini R, Erkoyuncu JA, Roy R, Torabmostaedi H (2018) A systematic review of augmented reality applications in maintenance. Robot Comput Integr Manuf 49:215–228. https://doi.org/10.1016/J.RCIM.2017.06.002
63. Tubis AA, Rohman J (2023) Intelligent warehouse in Industry 4.0—systematic literature review. Sensors 23:4105. https://doi.org/10.3390/s23084105
64. Tubis AA, Jodejko-Pietruczuk A, Nowakowski T (2024) Use of augmented reality as a tool to support cargo handling operations at the CARGO air terminal. Sensors 24:1099. https://doi.org/10.3390/s24041099
65. Wang X, Ong SK, Nee AYC (2016) Multi-modal augmented-reality assembly guidance based on bare-hand interface. Adv Eng Inform 30:406–421. https://doi.org/10.1016/j.aei.2016.05.004
66. Hou L, Wang X (2013) A study on the benefits of augmented reality in retaining working memory in assembly tasks: a focus on differences in gender. Autom Constr 32:38–45. https://doi.org/10.1016/j.autcon.2012.12.007
67. Guo A, Raghu S, Xie X, Ismail S, Luo X, Simoneau J, Gilliland S, Baumann H, Southern C, Starner T (2014) A comparison of order picking assisted by head-up display (HUD), cart-mounted display (CMD), light, and paper pick list. In: Proceedings of the 2014 ACM

international symposium on wearable computers. ACM, New York, NY, USA, pp 71–78. https://doi.org/10.1145/2634317.2634321
68. Throssel C. The benefits and challenges of augmented reality
69. Rejeb A, Keogh JG, Wamba SF, Treiblmaier H (2021) The potentials of augmented reality in supply chain management: a state-of-the-art review. Manage Rev Q 71:819–856. https://doi.org/10.1007/s11301-020-00201-w
70. Stoltz M-H, Giannikas V, McFarlane D, Strachan J, Um J, Srinivasan R (2017) Augmented reality in warehouse operations: opportunities and barriers. IFAC-PapersOnLine 50:12979–12984. https://doi.org/10.1016/j.ifacol.2017.08.1807
71. Sun R, Wu YJ, Cai Q (2019) The effect of a virtual reality learning environment on learners' spatial ability. Virtual Real 23:385–398. https://doi.org/10.1007/s10055-018-0355-2
72. Jensen L, Konradsen F (2018) A review of the use of virtual reality head-mounted displays in education and training. Educ Inf Technol (Dordr) 23:1515–1529. https://doi.org/10.1007/s10639-017-9676-0
73. Steuer J (1992) Defining virtual reality: dimensions determining telepresence. J Commun 42:73–93. https://doi.org/10.1111/j.1460-2466.1992.tb00812.x
74. Lee EA-L, Wong KW (2008) A review of using virtual reality for learning. In: Pan Z, Cheok AD, Müller W, El Rhalibi A (eds) Transactions on edutainment I. Lecture notes in computer science. Springer, Berlin, pp 231–241. https://doi.org/10.1007/978-3-540-69744-2_18
75. Mandal S (2013) Brief introduction of virtual reality & its challenges. Int J Sci Eng Res 4:304–309
76. Naranjo JE, Sanchez DG, Robalino-Lopez A, Robalino-Lopez P, Alarcon-Ortiz A, Garcia MV (2020) A scoping review on virtual reality-based industrial training. Appl Sci 10:8224. https://doi.org/10.3390/app10228224
77. Rupasinghe TD, Kurz ME, Washburn C, Gramopadhye AK (2011) Virtual reality training integrated curriculum: an aircraft maintenance technology (AMT) education perspective. education perspective. Int J Eng Educ 27:778–788
78. Howard MC, Gutworth MB, Jacobs RR (2021) A meta-analysis of virtual reality training programs. Comput Human Behav 121:106808. https://doi.org/10.1016/j.chb.2021.106808
79. Xie B, Liu H, Alghofaili R, Zhang Y, Jiang Y, Lobo FD, Li C, Li W, Huang H, Akdere M, Mousas C, Yu L-F (2021) A review on virtual reality skill training applications. Front Virtual Real 2. https://doi.org/10.3389/frvir.2021.645153
80. Mao RQ, Lan L, Kay J, Lohre R, Ayeni OR, Goel DP, de Darren SA (2021) Immersive virtual reality for surgical training: a systematic review. J Surg Res 268:40–58. https://doi.org/10.1016/J.JSS.2021.06.045
81. Tubis AA, Kierzkowski AA, Wolniewicz Ł, Mardeusz E, Restel FJ, Kisiel T, Zając M (2024) Risk assessment for the preparation of training scenarios for tram drivers. In: Prentkovskis O, Yatskiv (Jackiva) I, Skačkauskas P, Karpenko M, Stosiak M (eds) TRANSBALTICA XIV: Transportation science and technology. TRANSBALTICA 2023. Lecture notes in intelligent transportation and infrastructure. Springer, pp 337–348. https://doi.org/10.1007/978-3-031-52652-7_33
82. Tubis AA, Poturaj H, Mardeusz E, Kisiel T (2024) The use of fault tree analysis to create adverse event scenarios for the purposes of cargo zone crew training at the airport. In: Prentkovskis O, Yatskiv (Jackiva) I, Skačkauskas P, Karpenko M, Stosiak M (eds) TRANSBALTICA XIV: Transportation science and technology. TRANSBALTICA 2023. Lecture notes in intelligent transportation and infrastructure. Springer, pp 272–286. https://doi.org/10.1007/978-3-031-52652-7_27
83. Zangiacomi A, Pessot E, Fornasiero R, Bertetti M, Sacco M (2020) Moving towards digitalization: a multiple case study in manufacturing. Prod Plann Control 31:143–157. https://doi.org/10.1080/09537287.2019.1631468
84. Brennen JS, Kreiss D (2016) Digitalization. In: The international encyclopedia of communication theory and philosophy. Wiley, pp 1–11. https://doi.org/10.1002/9781118766804.wbiect111

85. Abdallah YO, Shehab E, Al-Ashaab A (2021) Understanding digital transformation in the manufacturing industry: a systematic literature review and future trends. Prod Manage Dev 19:e20200021. https://doi.org/10.4322/pmd.2021.001
86. Büyüközkan G, Göçer F (2018) Digital supply chain: literature review and a proposed framework for future research. Comput Ind 97:157–177. https://doi.org/10.1016/j.compind.2018.02.010
87. Tubis AA, Grzybowska K (2022) In search of Industry 4.0 and Logistics 4.0 in small-medium enterprises—a State of the Art review. Energies (Basel) 15:8595. https://doi.org/10.3390/en15228595
88. Abdallah YO, Shehab E, Al-Ashaab A (2022) Developing a digital transformation process in the manufacturing sector: Egyptian case study. IseB 20:613–630. https://doi.org/10.1007/s10257-022-00558-3
89. Genzorova T, Corejova T, Stalmasekova N (2019) How digital transformation can influence business model, case study for transport industry. Transp Res Procedia 40:1053–1058. https://doi.org/10.1016/j.trpro.2019.07.147
90. Tubis AA (2023) Digital maturity assessment model for the organizational and process dimensions. Sustainability 15:15122. https://doi.org/10.3390/su152015122
91. Mittal S, Romero D, Wuest T (2018) Towards a smart manufacturing toolkit for SMEs. In: Chiabert P, Bouras A, Noël F, Ríos J (eds) Product lifecycle management to support Industry 4.0. PLM 2018. IFIP advances in information and communication technology. Springer, pp 476–487. https://doi.org/10.1007/978-3-030-01614-2_44
92. Chen Q, Adey BT, Haas CT, Hall DM (2022) Exploiting digitalization for the coordination of required changes to improve engineer-to-order materials flow management. Constr Innov 22:76–100. https://doi.org/10.1108/CI-03-2020-0039
93. Belli L, Davoli L, Medioli A, Marchini PL, Ferrari G (2019) Toward Industry 4.0 with IoT: optimizing business processes in an evolving manufacturing factory. Front ICT 6. https://doi.org/10.3389/fict.2019.00017
94. Lipsmeier A, Kühn A, Joppen R, Dumitrescu R (2020) Process for the development of a digital strategy. Procedia CIRP 88:173–178. https://doi.org/10.1016/j.procir.2020.05.031
95. Nikmehr B, Hosseini MR, Martek I, Zavadskas EK, Antucheviciene J (2021) Digitalization as a strategic means of achieving sustainable efficiencies in construction management: a critical review. Sustainability 13:5040. https://doi.org/10.3390/su13095040
96. Ho WR, Tsolakis N, Dawes T, Dora M, Kumar M (2022) A digital strategy development framework for supply chains. IEEE Trans Eng Manag:1–14. https://doi.org/10.1109/TEM.2021.3131605
97. Ehie I, Ferreira LMDF (2019) Conceptual development of supply chain digitalization framework. IFAC-PapersOnLine 52:2338–2342. https://doi.org/10.1016/J.IFACOL.2019.11.555
98. Tubis AA, Grzybowska K, Król B (2023) Supply chain in the digital age: a scientometric-thematic literature review. Sustainability 15:11391. https://doi.org/10.3390/su151411391
99. Stank T, Esper T, Goldsby TJ, Zinn W, Autry C (2019) Toward a digitally dominant paradigm for twenty-first century supply chain scholarship. Int J Phys Distrib Logist Manag 49:956–971. https://doi.org/10.1108/IJPDLM-03-2019-0076
100. Khan M, Khan M, Ali A, Khan MI, Ullah I, Iqbal M (2022) Digitalization for fast, fair, and safe humanitarian logistics. Logistics 6:31. https://doi.org/10.3390/logistics6020031
101. Lee KL, Azmi NAN, Hanaysha JR, Alzoubi HM, Alshurideh MT (2022) The effect of digital supply chain on organizational performance: an empirical study in Malaysia manufacturing industry. Uncertain Supply Chain Manage 10:495–510. https://doi.org/10.5267/j.uscm.2021.12.002
102. Lee YY, Falahat M (2019) The impact of digitalization and resources on gaining competitive advantage in international markets: mediating role of marketing, innovation and learning capabilities. Technol Innov Manage Rev 9:26–38. https://doi.org/10.22215/timreview/1281
103. Louw C, Nieuwenhuizen C (2020) Digitalization strategies for SMEs: A cost vs. skill approach for website development. African J Sci Technol Innov Dev 12:195–202. https://doi.org/10.1080/20421338.2019.1625591

104. Dethine B, Enjolras M, Monticolo D (2020) Digitalization and SMEs' export management: impacts on resources and capabilities. Technol Inno Manage Rev 10:18–34. https://doi.org/10.22215/timreview/1344
105. Ivanov D, Dolgui A, Sokolov B (2019) The impact of digital technology and Industry 4.0 on the ripple effect and supply chain risk analytics. Int J Prod Res 57:829–846. https://doi.org/10.1080/00207543.2018.1488086
106. Zouari D, Ruel S, Viale L (2021) Does digitalising the supply chain contribute to its resilience? Int J Phys Distrib Logist Manag 51:149–180. https://doi.org/10.1108/IJPDLM-01-2020-0038
107. Pettit TJ, Croxton KL, Fiksel J (2019) The evolution of resilience in supply chain management: a retrospective on ensuring supply chain resilience. J Bus Logist 40:56–65. https://doi.org/10.1111/jbl.12202
108. Xu X, Choi T-M (2021) Supply chain operations with online platforms under the cap-and-trade regulation: impacts of using blockchain technology. Transp Res E Logist Transp Rev 155:102491. https://doi.org/10.1016/j.tre.2021.102491
109. Kamali Saraji M, Streimikiene D, Kyriakopoulos GL (2021) Fermatean fuzzy CRITIC-COPRAS method for evaluating the challenges to Industry 4.0 adoption for a sustainable digital transformation. Sustainability 13:9577. https://doi.org/10.3390/su13179577
110. Ahi AA, Sinkovics N, Shildibekov Y, Sinkovics RR, Mehandjiev N (2022) Advanced technologies and international business: a multidisciplinary analysis of the literature. Int Bus Rev 31:101967. https://doi.org/10.1016/j.ibusrev.2021.101967
111. Anastasiadis F, Tsolakis N, Srai J (2018) Digital technologies towards resource efficiency in the agrifood sector: key challenges in developing countries. Sustainability 10:4850. https://doi.org/10.3390/su10124850
112. Heavin C, Power DJ (2018) Challenges for digital transformation—towards a conceptual decision support guide for managers. J Decis Syst 27:38–45. https://doi.org/10.1080/12460125.2018.1468697
113. Kokolek N, Jakovic B, Curlin T (2019) Digital knowledge and skills—key factors for digital transformation. Ann DAAAM Proc Int DAAAM Symp 30:46–53. https://doi.org/10.2507/30th.daaam.proceedings.006
114. Winkelhaus S, Grosse EH (2020) Logistics 4.0: a systematic review towards a new logistics system. Int J Prod Res 58:18–43. https://doi.org/10.1080/00207543.2019.1612964
115. Bigliardi B, Filippelli S, Petroni A, Tagliente L (2022) The digitalization of supply chain: a review. Procedia Comput Sci 200:1806–1815. https://doi.org/10.1016/J.PROCS.2022.01.381
116. Wang K (2016) Logistics 4.0 solution-new challenges and opportunities, 68–74. https://doi.org/10.2991/iwama-16.2016.13
117. Facchini F, Oleśków-Szłapka J, Ranieri L, Urbinati A (2019) A maturity model for logistics 4.0: an empirical analysis and a roadmap for future research. Sustainability 12:86. https://doi.org/10.3390/su12010086
118. Szymańska O, Adamczak M, Cyplik P (2017) Logistics 4.0—a new paradigm or set of known solutions? Res Logis Prod 7:299–310. https://doi.org/10.21008/j.2083-4950.2017.7.4.2
119. Wawrla L, Maghazei O, Netland T (2019) Applications of drones in warehouse operations. Whitepaper
120. Hu L, Xiang C, Qi C (2020) Research on traceability of cold chain logistics based on RFID and EPC. IOP Conf Ser Mater Sci Eng 790:012167. https://doi.org/10.1088/1757-899X/790/1/012167
121. Nikola G, Schulz K, Ivanochko I (2021) Use of RFID technology in retail supply chain. In: Kryvinska N, Greguš M (eds) Developments in information & knowledge management for business applications. Studies in systems, decision and control. Springer, pp 555–587. https://doi.org/10.1007/978-3-030-62151-3_14
122. Caccami MC, Amendola S, Occhiuzzi C (2019) Method and system for reading RFID tags embedded into tires on conveyors. In: 2019 IEEE international conference on RFID technology and applications (RFID-TA). IEEE, pp 141–144. https://doi.org/10.1109/RFID-TA.2019.8892245

123. Ren Y, Li H (2018) Building materials management system based on RFID technology. Int J RF Technol 9:63–74. https://doi.org/10.3233/RFT-180133
124. Landmark AD, Sjøbakk B (2017) Tracking customer behaviour in fashion retail using RFID. Int J Retail Distrib Manage 45:844–858. https://doi.org/10.1108/IJRDM-10-2016-0174
125. Cheng C-H, Kuo Y-H, Lam H, Petering M (2021) Real-time location-positioning technologies for managing cart operations at a distribution facility. Appl Sci 11:4049. https://doi.org/10.3390/app11094049
126. Casella G, Bigliardi B, Bottani E (2022) The evolution of RFID technology in the logistics field: a review. Procedia Comput Sci 200:1582–1592. https://doi.org/10.1016/j.procs.2022.01.359
127. Chanchaichujit J, Balasubramanian S, Charmaine NSM (2020) A systematic literature review on the benefit-drivers of RFID implementation in supply chains and its impact on organizational competitive advantage. Cog Bus Manage 7:1818408. https://doi.org/10.1080/23311975.2020.1818408
128. Hodgkinson D, Johnston R. Aviation law and drones: unmanned aircraft and the future of aviation. Routledge
129. Singhal G, Bansod B, Mathew L (2018) Unmanned aerial vehicle classification, applications and challenges: a review. Preprints (Basel)
130. Yoo W, Yu E, Jung J (2018) Drone delivery: Factors affecting the public's attitude and intention to adopt. Telematics Inform 35:1687–1700. https://doi.org/10.1016/j.tele.2018.04.014
131. Fernández-Caramés TM, Blanco-Novoa O, Froiz-Míguez I, Fraga-Lamas P (2019) Towards an autonomous Industry 4.0 warehouse: a UAV and blockchain-based system for inventory and traceability applications in Big Data-driven supply chain management. Sensors (Basel) 19. https://doi.org/10.3390/s19102394
132. Kostrzewski M, Varjan P, Gnap J (2020) Solutions dedicated to internal logistics 4.0. In: Grzybowska K, Awasthi A, Sawhney R (eds) Sustainable logistics and production in Industry 4.0. EcoProduction. Springer, pp 243–262. https://doi.org/10.1007/978-3-030-33369-0_14
133. Tubis AA, Ryczyński J, Żurek A (2021) Risk assessment for the use of drones in warehouse operations in the first phase of introducing the service to the market. Sensors 21:6713. https://doi.org/10.3390/s21206713
134. Maghazei O, Netland T (2019) Drones in manufacturing: exploring opportunities for research and practice. J Manuf Technol Manag 31:1237–1259. https://doi.org/10.1108/JMTM-03-2019-0099
135. Tubis AA, Żurek A (2022) The use of drones to handle internal logistic flows in the company from the chemical sector. In: Conference proceedings—CLC 2022, June 15–17, pp 107–112, Bojnice. https://doi.org/10.37904/clc.2022.4572
136. Tubis AA, Żurek A (2021) Adverse event analysis in the application of drones supporting safety and identification of products in warehouse storage operations. In: Proceedings of the 31st European safety and reliability conference (ESREL 2021). Research Publishing Services, Singapore, pp 2533–2539. https://doi.org/10.3850/978-981-18-2016-8_446-cd
137. Tubis AA, Poturaj H (2022) Risk related to AGV systems—open-access literature review. Energies (Basel) 15:8910. https://doi.org/10.3390/en15238910
138. Cramer M, Cramer J, De Schepper D, Aerts P, Kellens K, Demeester E (2019) Benchmarking low-cost inertial measurement units for indoor localisation and navigation of AGVs. Procedia CIRP 86:204–209. https://doi.org/10.1016/j.procir.2020.01.044
139. Peterson SB, Michalek JJ (2013) Cost-effectiveness of plug-in hybrid electric vehicle battery capacity and charging infrastructure investment for reducing US gasoline consumption. Energy Policy 52:429–438. https://doi.org/10.1016/j.enpol.2012.09.059
140. Bechtsis D, Tsolakis N, Vlachos D, Iakovou E (2017) Sustainable supply chain management in the digitalisation era: the impact of automated guided vehicles. J Clean Prod 142:3970–3984. https://doi.org/10.1016/J.JCLEPRO.2016.10.057
141. De Ryck M, Versteyhe M, Debrouwere F (2020) Automated guided vehicle systems, state-of-the-art control algorithms and techniques. J Manuf Syst 54:152–173. https://doi.org/10.1016/J.JMSY.2019.12.002

142. Żuchowski W (2022) The smart warehouse trend: actual level of technology availability. Logforum 18:227–235. https://doi.org/10.17270/J.LOG.2022.702
143. Le-Anh T, De Koster MBM (2006) A review of design and control of automated guided vehicle systems. Eur J Oper Res 171:1–23. https://doi.org/10.1016/J.EJOR.2005.01.036

Chapter 4
Risk Analysis in Internal Transport Systems in Logistics 4.0

Contents

According to the characteristics presented in Chap. 3, among the mobile solutions of Industry 4.0 supporting internal transport, AGV systems and drones are the most common. Their implementation is related to the need to improve the efficiency of internal transport processes and to reduce the disruptions occurring in the process,

A. Tubis, *The New Paradigm of Risk in Internal Transport Supporting Logistics 4.0 System*, Lecture Notes in Intelligent Transportation and Infrastructure,
https://doi.org/10.1007/978-3-031-64920-2_4

the source of which is the human factor. Implementing these mobile solutions thus responds to the need to minimise the risks present in traditional internal logistics systems. However, at the same time, these systems generate new types of adverse events characteristic of cyber-physical systems. To confirm the research gap defined in Chap. 1, the Author and her team conducted a comprehensive literature review to identify current research trends related to risks in internal transport processes supported by Logistics 4.0 solutions. A summary of this research will be presented in this chapter.

4.1 Risks Concerning the Operation of the AGV System in Internal Transport

Chapter 3 presents the characteristics of AGV systems, as well as the challenges and benefits associated with their use in complex cyber-physical systems. AGV systems are among the key research areas conducted by the Author in recent years. This research primarily focuses on assessing the risks associated with the operation of AGV systems and their interaction with humans in a shared workspace. The research results have been published, among others, in [1–3].

One of the phases of the conducted research was a systematic literature review on the risks associated with the operation of AGV systems, which is described in detail in [1]. The literature review looked at articles and proceeding papers from 2018 to 2022 published by three publishers: Elsevier, IEE and MDPI. These publishers have had the most significant publication output in this area in recent years. A total of 155 articles and 44 proceedings papers were accepted for the analysis, identified from a search conducted according to the PRISMA method based on the searches presented in Table 4.1.

The analysis of the publications made it possible to identify 12 research trends related to the risks of operating AGVs. These trends are shown in Fig. 4.1. Based on the critical literature analysis, all 199 publications were classified into the highlighted thematic categories. However, for the research presented in this book, the Author's attention will only focus on selected articles from this review.

Table 4.1 Search query for publications on risk in AGV systems

TITLE_ABS_KEY ("AGV" OR "automated guided vehicle" AND "risk") AND LANGUAGE (English) AND PUBYEAR > 2017

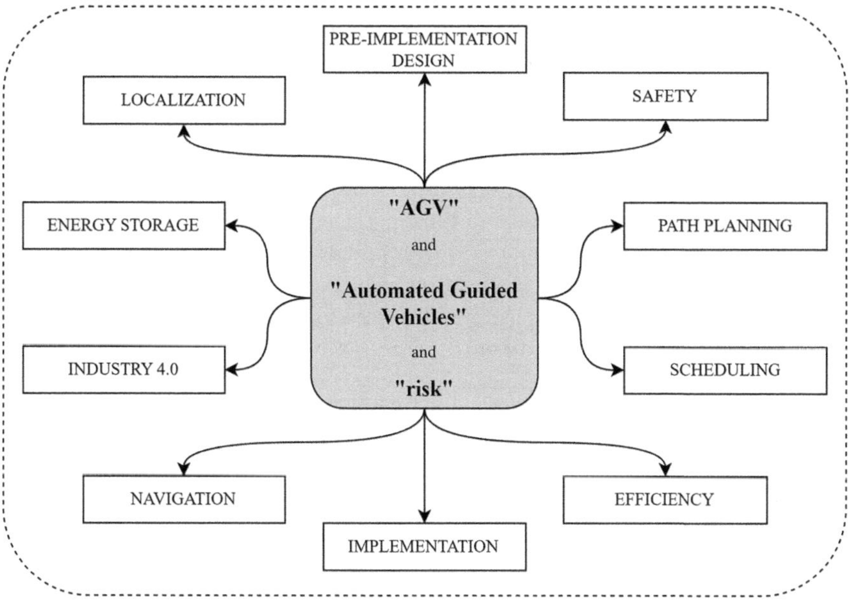

Fig. 4.1 Main research trends related to AGV system risk (based on [1])

4.1.1 Safety

Safety has to be considered the most critical research trend relating to the risks associated with operating an AGV system. This is obviously because most risk analyses relating to automated or autonomous systems focus primarily on safety. The issues of the safe movement of these devices through logistics facilities, including shared spaces with operational staff, significantly determine other research trends relating to vehicle navigation, path planning, localisation and control systems. At the same time, the Author distinguished three dominant research directions within this category, as shown in Fig. 4.2.

The first research area identified is the analysis of the causes of hazardous situations related to the operation of AGVs. This area can be divided into two groups: (a) hazardous situations involving only AGVs and (b) hazardous situations involving AGVs and other objects. Within the first group, ongoing research focuses on analysing collisions between vehicles during their missions [4, 5]. Therefore, the researchers focus on aspects concerning the crossing of the paths of the moving vehicles and the coordination of the work of the trucks within the schedules created. This is because both research aspects were identified as the leading causes of vehicle collisions and jams. The second group of hazardous situations refers to collisions

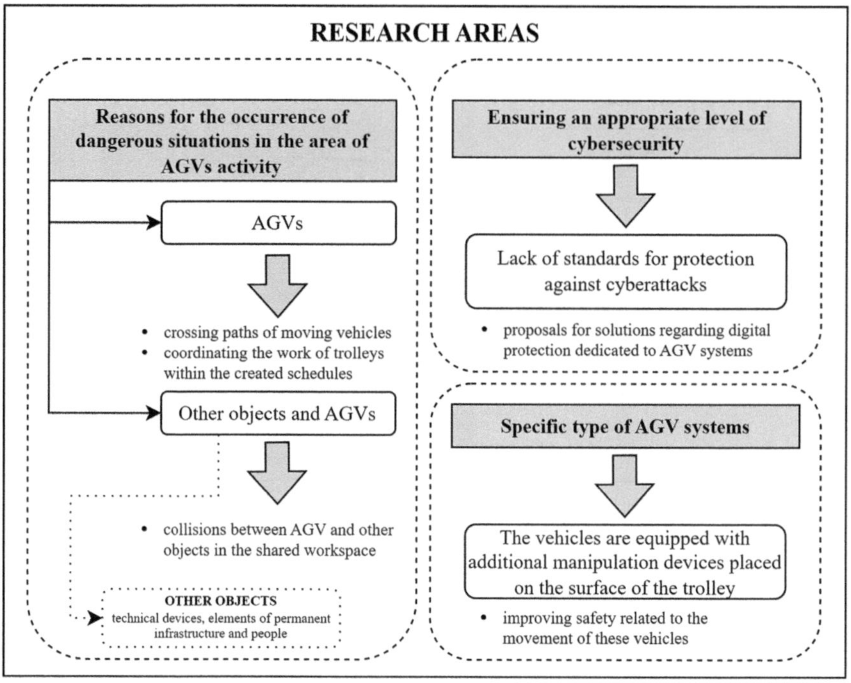

Fig. 4.2 Research directions in the "Safety" category

between AGVs and other objects in the shared workspace [6]. These objects usually include other technical equipment, fixed infrastructure elements and people. Therefore, authors analysing such events have proposed solutions aimed at improving navigation [7], path planning [8], and obstacle-braking systems [9].

The second area concerning the safe operation of AGV systems was issues related to ensuring an adequate level of cyber security. The authors of the analysed publications mainly drew attention to the lack of standards for protection against cyberattacks, which would consider the specific functioning of AGV systems [10]. Many researchers also proposed digital protection solutions dedicated to AGV systems [11].

The third research area referred to a specific type of AGV system in which vehicles are equipped with additional manipulation devices placed on the trolley surface. As the vehicle moves, the additional equipment introduced into the tremor can cause the whole device to lose stability. Therefore, this group includes publications in which the authors analyse the safety enhancement issues associated with the movement of these vehicles [12, 13].

4.1.2 Path Planning

The reason for many adverse events involving AGVs is the difficulty of planning how the vehicles move through the operational zone. For this reason, the second major research trend is path planning. For this group, the majority of risk assessment research has focused primarily on developing vehicle movement algorithms to avoid collisions and conflicts between AGVs during their missions [8, 14, 15], as well as to reduce their collisions with other objects in the AGVs' movement space [16, 17]. In some studies, the risk analysis results allowed the type and probability of obstacles along the vehicle's route to be determined so that their occurrence was considered when optimising routes [18]. A solution to emerging adverse events during the movement of AGVs is dynamic path planning. Research aimed at reducing the occurring risk of collisions and conflicts in this way is presented in [19], among others.

4.1.3 Navigation

The second critical cause of adverse events occurring is interference with vehicle navigation. AGV systems can be navigated in two ways:

- with predefined paths that use various human-installed signage systems (lines, QR codes, reflections);
- based on sensors and systems that allow vehicles to "observe" the operational environment (e.g. LiDAR, cameras, multi-antennas).

Both navigation systems have their drawbacks and limitations, which are the focus of the risk analyses. The results of the analyses carried out are the basis for

- preparing proposals for new solutions [20];
- improvements to current systems [21];
- identification of risks associated with implemented solutions [22];
- identification of risks arising from environmental conditions [23].

4.1.4 Localisation

Risks associated with vehicle navigation are often linked to interference with vehicle location (e.g., lack of information about the current vehicle location and erroneous vehicle location parameters reported by the system). For this reason, another distinguished research category related to the risks associated with AGV systems is "Localisation". This is because the efficient navigation of an AGV system requires the provision of real-time, up-to-date information on the vehicle's location. Therefore, the level of risk involved in moving vehicles strongly depends on the system meeting the requirements for the accuracy of the current location information provided. Thus, research in this category focuses on improving localisation methods or adapting them to increasingly demanding operating conditions [24–26].

4.1.5 Scheduling

Issues concerning the risk assessment of the operation of AGVs are often related to the stage of planning their work and assigning them missions to accomplish. For this reason, one of the more numerous categories highlighted in this review is "Scheduling". When creating the scheduling of an AGV system, it is crucial to consider the risks associated with limited company resources [27]. Most publications in this group have focused on the analysis of Flexible Production Systems (EPS), where the high dynamics of change strongly affect the risk of disruption to the material supply process handled by the AGV system [28]. Therefore, EPS requires the implementation of dynamic vehicle scheduling based on learning algorithms [15, 29] or using a digital twin to refine the system's work scheduling [30]. The dynamic scheduling implemented also aims to eliminate disruptions due to changes in the production process [15, 31].

As noted under the Safety category, an important aspect of scheduling an AGV system is considering the risk of collisions between vehicles simultaneously performing their assigned tasks. For this reason, several publications analyse different scheduling methods to minimise collisions between moving vehicles [32, 33].

Another critical research issue is the risk associated with the power system of AGVs. Therefore, ongoing research is directed towards optimising the assignment of missions to be carried out, considering the capacity of the battery powering the vehicle [34, 35] or the potential savings of the battery consumed by the vehicle during the mission [36, 37].

4.1.6 Power Supply For AGVs

The issue of power supply for the AGV system was highlighted as a separate research category. This category includes publications dealing not only with the capacity of the battery supplying the vehicle but also research to select an appropriate power supply system. The risk analysis in these publications is used to assess adverse events related to:

- the implementation of a proposed improved or new power system [38, 39];
- the justification of the need to change the power supply system due to risks in the existing system [40].

Particular attention of researchers is also focused on solutions that allow vehicles to be powered up without human intervention, thus reducing human error and dependence on the availability of operational staff (elimination of wastage associated with waiting for service). Therefore, published research looks at wireless [41, 42] or inductive power transmission [43].

4.1.7 Industry 4.0

AGV systems are a solution whose growing popularity is linked to the development of Industry 4.0. The I4.0 concept introduces new challenges to the operation of an AGV system. The results of the risk assessment carried out, in this case, may refer to the need to meet specific requirements for an AGV system operating in an I4.0 environment [44, 45] or to adapt the AGV system to the needs of SME companies implementing I4.0 solutions [46, 47].

An important I4.0 solution to support the management of risks associated with the operation of AGV systems is digital twins. This digital solution is used to:

- improve AGV system navigation [48];
- improve the scheduling of the AGV system [30];
- controlling the movement of vehicles in their workspace [49];
- improving the management process of AGV systems in general [50].

4.1.8 Control System

The "Control system" category refers to two functions: (a) monitoring the correct functioning of the AGV system and (b) controlling vehicles in the event of a disturbance. Risk assessment results in monitoring the AGV system's correct operation are based on measurements made using information systems [51, 52] and various sensors [53]. Some publications refer to using the control system to reduce interference occurring during the execution of missions by vehicles [54].

4.1.9 Implementation in Specific Systems

An interesting category is represented by publications in which the assessed risks are related to using AGVs in non-standard anthropotechnical systems. In these studies, the risk assessment is carried out to identify the difficulties encountered in their implementation. Addressing these challenges during the system design and planning stages can reduce the frequency of adverse events. Noteworthy among the results described in this category are studies on the application of AGV systems in:

- smart parking systems in large cities [55];
- large greenhouse installations [56];

- the car-loading process of Ro-Ro ships [57];
- routing in the mountains [58].

4.2 Cooperation between Human and AGV System

The results of the Author's research indicated that most of the publications on AGV systems that addressed risk issues referred to safety as the leading research category. Most of the publications included in the other categories also directly or indirectly referred to these aspects. It is worth noting, however, that only a few publications included humans. Of course, people appear in all these safety studies. However, they are only potential recipients of the effects of the adverse events or one of the possible objects with which a vehicle may collide. Meanwhile, these are not the only adverse events in which humans appear as a source of risk or a victim of its potential effects. Therefore, the Author and her research team felt that this was a significant research gap that needed to be acknowledged by detailing previous research and focusing attention exclusively on issues concerning the interaction of AGV systems with humans.

Therefore, the second research phase concerned a systematic literature review already focused exclusively on the collaboration between humans and AGV systems. Here, the PRISMA method was also used to search for publications that met the conditions defined according to the queries presented in Table 4.2.

The two largest databases, i.e. Web of Science and Scopus, were used to search for publications related to the research question under analysis. Some publications were duplicated, but this was corrected when sorting out the material for study. After the selection process, 117 publications were accepted for analysis, including 81 proceedings papers and 36 articles published between 2018 and 2023 (excluding December 2023). Detailed results from the analysis of the full review of the indicated 177 publications are presented in [3]. However, only the main research trends and general conclusions from the literature review will be discussed in this book.

Publishers that publish the most research results on human–machine collaboration include IEE, Springer and Elsevier. The analysis of the issues discussed in these publications identified three dominant research trends, as shown in Fig. 4.3.

Table 4.2 Search query for publications on human-AGV system collaboration

TITLE_ABS_KEY ("AGV" OR "automated guided vehicle" OR "autonomous guided vehicle" AND "human") AND LANGUAGE (English) AND PUBYEAR > 2017

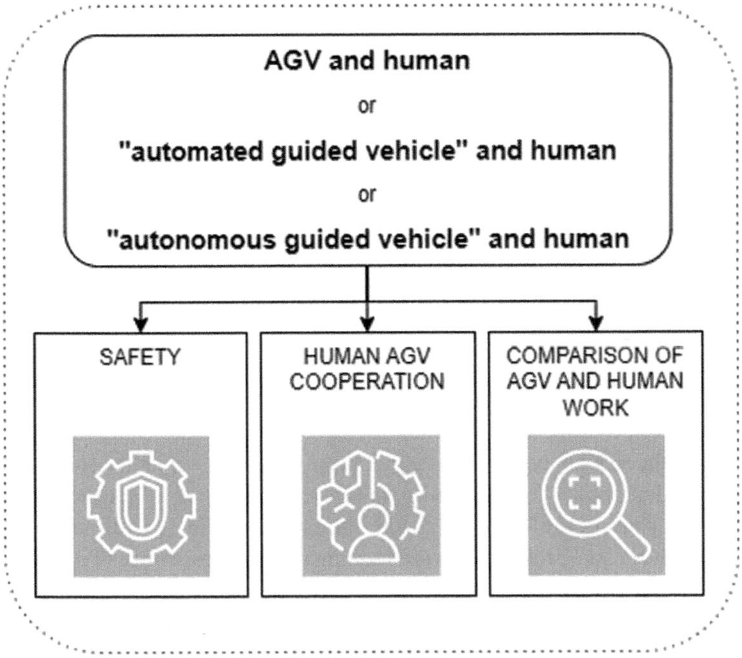

Fig. 4.3 Main research trends in publications on man and AGVs (based on [3])

4.2.1 Designing a Safe Work Environment

As with the first literature review, safety issues are an important research category. For the publications reviewed, the main focus of the researchers was on issues concerning the design of a safe work environment. Technical systems working with humans must not endanger their health or life, so the risk assessment mainly emphasises identifying hazardous situations. Accordingly, three lines of research have been identified in this category, as shown in Fig. 4.4.

The first research direction is the analysis of human (obstacle) detection by AGV systems. Vehicles in operation should obtain the required real-time information on the location of people in the space in which the vehicles are moving. Therefore, to reduce the risks associated with collisions between AGVs and humans, the authors of the publication investigated the possibility of using different data collection and human location technologies, such as:

- cameras [59, 60];
- laser sensors [61, 62];
- vehicles [59, 63];
- ultrasound [61];
- human-located sensors (Ultra-wideband system) [64, 65].

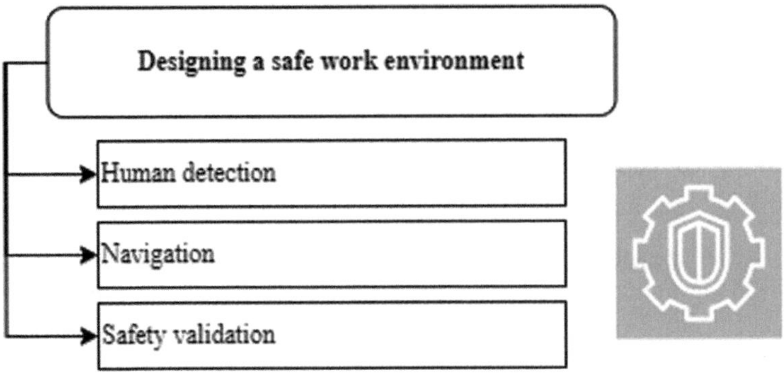

Fig. 4.4 Research directions in the category "Designing a safe work environment" (based on [3])

However, the risk analyses indicate that these solutions may have specific limitations, which determine the detection system's reliability and affect the safety level of the human sharing the workspace with the AGV system. Therefore, to mitigate the risks involved, some researchers are looking into the possibility of implementing hybrid detection systems that make it possible to increase safety even in a dynamic working environment [7, 66].

The second group of safety-related publications deals with aspects of AGV system navigation, which is also confirmed by the results of the first literature review. In the publications analysed, the authors focused mainly on the correct navigation of vehicles along established paths so as not to cause conflicts or collisions with people [67–69]. Also relevant to this research issue is how a vehicle undertakes a mission after an emergency stop caused by the appearance of an obstacle (human) on the road. In this case, the researchers' attention is directed to the issues of autonomous or guided resumption and the determination of a new path to avoiding the obstacle [70–72].

The third group consisted of publications whose results are oriented towards safety validation. This validation can refer to three issues:

(1) The feasibility of using risk analysis to validate the safety associated with the operation of an AGV system.

Research on such analyses has been reported in [73, 74]. The results obtained in [73] made it possible to identify faulty elements (areas) in an AGV system and, on this basis, to assess the risk of situations posing a threat to its operation. In contrast, Anastasi et al. [74] focused on risks relating to human-vehicle interaction.

(2) The use of simulation tools and testing programmes to improve the safety of human-AGV interaction. Research in this group has addressed:

   a. Testing methods of human-AGV communication [75];
   b. Testing vehicle positioning methods [76];
   c. Testing a pick-up and delivery system in which a human interacts with the AGV [77];

d. Creating a virtual environment to simulate AGV operation with human safety in mind [78].

(3) Safety assessment for accident risk prevention.

Research in this area relates to implementing human-vehicle accident prevention systems [79] or systems to manage disturbance [80].

4.2.2 Human-AGV Cooperation

The second identified research category shown in Fig. 4.3 is human-AGV cooperation. Again, three research areas are identified, as shown in Fig. 4.5.

An important research issue is to place the human at the centre of attention when designing the AGV system and planning its operation. To achieve this, information about the behaviour and modus operandi of the people interacting with the AGV system must be taken into account first and foremost. Bergman et al. [81] and Locking [82] designed their operation based on human behaviour and nature to increase the predictability and unambiguity of operations performed by AGVs. Bach et al. [83] even developed a controller for the AGV system that mimics human driving behaviour, such as adjusting speed when cornering. Prati et al. [84] developed a set of guidelines for communication and data exchange between humans and mobile robots.

Another critical aspect of human–machine collaboration is mutual communication. Therefore, it is crucial to develop clear and convenient human interfaces to exchange messages between humans and vehicles. However, Vlachos [85] emphasises that the critical issue is not only a tool in the form of an interface adapted to the nature of the collaboration but also raising awareness of the need for human-vehicle interaction. David [86] additionally points out that the human-AGV communication system should be interoperable and adapted to different types of vehicles. When

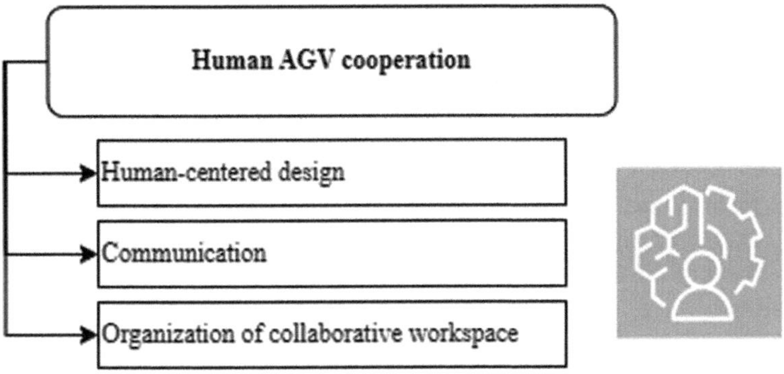

Fig. 4.5 Research directions in the category "Human-AGV cooperation" (based on [3])

specifying the requirements for a human-AGV communication panel, it is important to remember to meet the requirements for proper ergonomics, particularly issues such as panel location height, tilt angle, and operator distance [87].

The data collection system by the AGV system and its transmission to the workers are also critical issues. Ballal et al. [88] proposed a wireless data collection and sharing system. The sensor system continuously collects defined data as the vehicles move, which is then transmitted via Wi-Fi and displayed to the worker on a laptop.

Many researchers stress that human-vehicle communication should be two-way and real-time [72, 84]. Various solutions to improve human-AGV communication are described in the literature. Among the most popular are:

- AGV follow-me system [89, 90];
- AGV call system [91, 92];
- AGV guide-me system [93];
- AGV control system primarily uses hand gestures for control [94, 95].

As described in scientific publications, the final stage of human-AGV collaboration is the optimal design of the shared workspace. Coelho et al. [96] point to the need for some balance in numbers between human resources and AGVs in a shared workspace. An area of much research is primarily the order-picking process and the appropriate coordination of people and AGVs in planned activities [97–99]. In some studies, AR technology can further support collaboration in a shared workspace. An example of such research is presented in [72], in which an AR environment was developed to track vehicle routes. Users can also indicate areas where only people are present, spaces where only AGVs move, and zones shared by people and vehicles.

4.2.3 Comparison of Agv and Human Work

The third research category identified is the comparison of human and AGV system operation. The comparative analyses prepared are usually used to formulate arguments for choosing a particular operating system. Three dominant research trends were identified in this category, as shown in Fig. 4.6.

Most research focuses on the effects of human labour and AGV systems in labour-intensive, time-consuming, yet repetitive processes [100, 101]. Indeed, many researchers note that human labour is very costly, so using it to perform simple, repetitive operations is economically unviable [102]. An example of such a process is order picking, and examples of comparisons for this area are described in [101], among others. A comparison of human labour and the AGV system is the subject of studies that aim to demonstrate the benefits of implementing automated solutions in the form of:

- increased productivity [100, 103–105];
- increase in efficiency [100, 102, 103, 105–110];
- time reduction [104, 111];

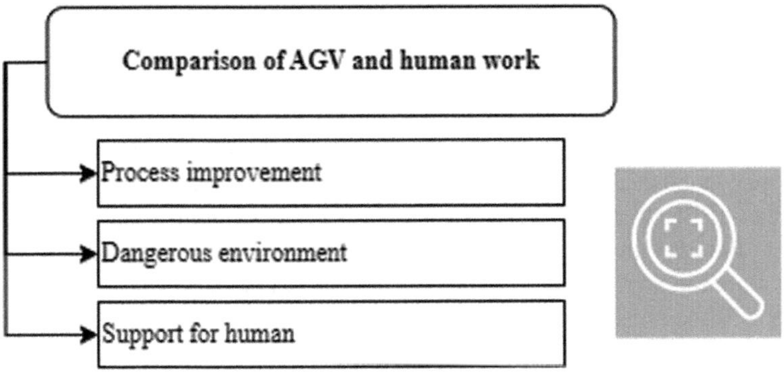

Fig. 4.6 Research trends in the category "Comparison of AGV and human work" (based on [3])

- reduction of human errors [103, 106, 109];
- congestion [105];
- improvement of safety [103, 105, 107, 110].

AGV systems are also implemented to assist or replace humans in tasks carried out in hazardous working environments. Research in this area was carried out primarily during the COVID-19 pandemic, and the benchmarking studies conducted indicated the enormous potential of AGV systems to replace humans in deliveries in contaminated areas, patient care and disinfection processes [112, 113]. Replacing humans with AGV systems for these processes reduced the exposure of healthcare personnel to extreme infections and fatal illnesses.

A second area of application of AGV systems instead of humans is in operations that physically burden staff, such as transporting sick people between hospital rooms [114] or transporting dirty dishes or heavy food in restaurants [104].

4.3 Risk Related to the Use of Drones in the Warehouse

A second Industry 4.0 solution to support the material handling process is drones, the characteristics of which are also presented in Chap. 3. Using drones in last-mile delivery logistics has attracted much interest recently. However, current legislation in many European Union countries means that there are currently many restrictions on the use of UAVs in external transport, and this situation is unlikely to change in the coming years. For this reason, many researchers and companies involved in this technology focus on the potential for developing drone applications, primarily in internal logistics. However, implementing these solutions in indoor facilities, for processes carried out in a workspace shared with people, generates many new risks specific to this technology.

Following in the footsteps of the research presented in Sect. 4.1, the same study was conducted by the Author and her team concerning the risks associated with the operation of UAV systems. A systematic literature review carried out according to the guidelines of the PRISMA method, was used to identify research trends on the topic under study. The search for publications was based on the Web of Science database and the searches presented in Table 4.3.

The publications from such a defined search were limited to articles and proceedings papers published by the three publishers with the largest share of publications in this area—IEEE (Institute of Electrical and Electronics Engineers), MDPI and Elsevier. For the final analysis, 257 papers from 2019 to 2023 were accepted. Detailed results from the literature review are described in [115]. However, for this book, the Author will focus only on the most important issues relevant to the developed risk assessment method.

The literature review made it possible to distinguish eight leading research trends related to the risk of using UAV systems. These categories are presented in Fig. 4.7.

Table 4.3 Search query for publications on human collaboration with AGV systems

TITLE_ABS_KEY ("drone" AND "risk") AND LANGUAGE (English) AND PUBYEAR > 2018

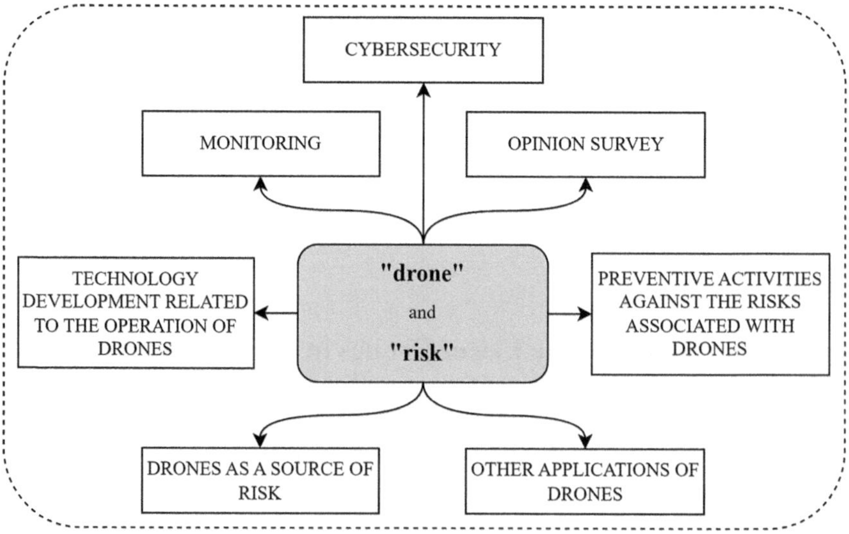

Fig. 4.7 Major research trends in drone risk publications (based on [115])

4.3.1 Monitoring

Most publications on risk and drones are concerned with monitoring selected phenomena. Drones are used extensively in monitoring processes because they can take pictures or collect measurements in difficult or dangerous conditions for humans. The most popular areas of drone use in monitoring, identified from the literature review conducted, are shown in Fig. 4.8.

The results of the analysis have shown that drone monitoring typically has two functions:

- Monitoring complements field-based surveys—first, human field surveys are carried out, and then the survey results are supplemented based on data from the drone missions.
- Monitoring results form the basis for the demand for surveys in a given area—phenomena that require field surveys by a research team are identified based on the data recorded by the drones.

However, it is worth noting that in many of the publications in this category, drones are only used to take photographs and measurements. The collected data is then processed and analysed by machine learning algorithms or artificial intelligence. Therefore, the results published in the papers concerned combining UAV systems with artificial intelligence, e.g. for monitoring natural disasters such as fire [116–118]

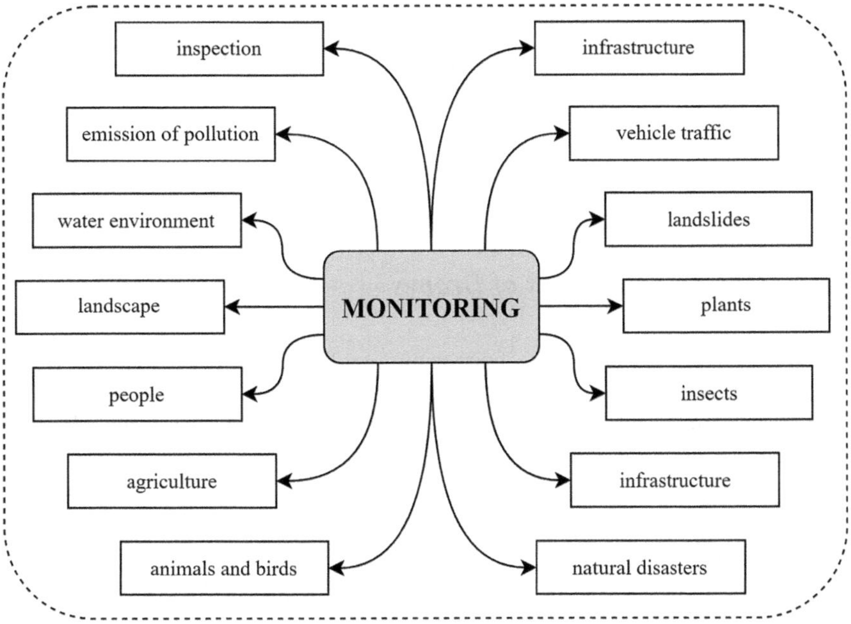

Fig. 4.8 Identified areas of drone use in monitoring processes (based on [115])

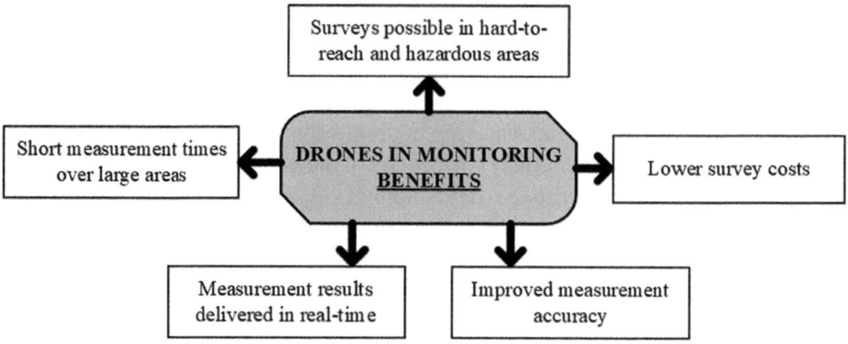

Fig. 4.9 Benefits of using drones in monitoring processes (based on [115])

or flooding [119–124]. These situations often require rapid decision-making based on the anticipated development of the phenomenon under investigation, which can save lives.

Drones also play an important role in various types of inspections, including but not limited to:

- monitoring the security of critical infrastructure [125];
- vigilance of construction works [126, 127];
- identification of hazardous phenomena (e.g. corrosion) concerning the condition of industrial structures or wind farms [128, 129];
- inspection of transport infrastructure [130, 131];
- detection of unexploded ordnance [132] and anti-personnel mines [133].

The significant benefits of their use determine the increasing popularity of incorporating drones into monitoring processes. The primary rationale for drone deployment in this area is shown in Fig. 4.9.

4.3.2 Other Applications of Drones

Monitoring is one of many areas of application for UAV systems. An equally important application area is the possibility of transporting small cargo with drones. This solution has many supporters primarily because of the benefits associated with their use, such as lower implementation costs, greater efficiency, and increased safety of the transport missions performed [134–136]. There are many opportunities for the commercial use of drones in transportation. Interesting examples of such use in Australia are presented in [137]. Drones are also used for delivery in the construction industry [138], the chemical sector [139, 140], and in agriculture for spreading materials [136]. Using drones to transport medical devices and blood while complying with regulations is also gaining popularity [141–145].

The ability to use AGV systems for transport and data collection makes this technology increasingly applicable to:

- search and rescue missions [146–153];
- disaster assessment and recovery [146, 147, 149, 154, 155];
- ground transport support in emergencies [141, 156, 157].

During the COVID-19 pandemic, AGV systems were used to transport medical supplies, especially in non-contact deliveries and disinfection of large areas [158]. Other possible areas of drone application described in the publications analysed include:

- warehouse inventories [159, 160];
- fighting mosquitoes [161] or desert locusts [162];
- exploration of other planets—Mars [163];
- rehabilitation of birds of prey [164];
- the herding of wild horses [165].

In all publications on the use of UAV systems for transport or data collection, the risk aspects are described in two ways:

- the risk assessment is about the use of drones in the study area or
- drones are a response to a risk that currently exists in the operation of a particular process

4.3.3 Technology Development Related to the Operation of Drones

Another critical research category is the direction of technology development related to the operation of drones. As noted at the beginning of this section, there are currently many industry constraints on the use of drones. An interesting analysis of technological challenges and regulatory constraints is presented in [141]. A critical technological challenge associated with the operation of UAV systems is ensuring adequate safety during their missions. Therefore, ongoing research is directed towards developing technologies to aid path planning, navigation and obstacle avoidance [146, 154, 166–172]. An important step in the operation of a drone is its landing. Therefore, many researchers focus on technological solutions enabling automatic landing [173–175], including on vertical surfaces [176].

Some studies address the risks of new technologies at the design stage and design changes in drone construction [177–179]. Studies also raise issues about technological solutions to improve drone communication [180], including human-drone interaction via gestures [181].

4.3.4 Drones as a Source of Risk

Much of the research on UAV systems focuses primarily on the benefits of drone use in various areas. However, due to the choice of keywords in the search process, one of the trends identified is research on the impact of drones as a source of risk present. These analyses often concern the risk of drones colliding with other objects—humans [182] or machines [183]. The source of such risk is primarily drones controlled by amateurs, whose lack of knowledge and experience can cause collisions, resulting in loss of human health or life and damage to infrastructure [184]. Examples of hazardous situations in which a drone is a source of risk concern:

- damage to the drone [185];
- execution of a mission by a drone in a space shared by people [186, 187];
- execution of missions near an airport [183, 188].

The growing popularity of using UAV systems to monitor flora and fauna has led many researchers to analyse the impact and reactions of animals [189] and birds [190–192] to drones appearing in their close vicinity. Indeed, drones are a source of noise and visual pollution that can interfere with the normal functioning of living organisms (including humans) [193].

4.3.5 Cybersecurity

Drones are a cyber-physical system and, like other I4.0 solutions, are vulnerable to cyber-attacks. Research results presented in publications indicate that small commercial drones [194] and large drone swarms [195] are at risk of attacks. The main reason for successful cyber-attacks on UAV systems is the system's vulnerabilities, exploited by cyber-terrorists. For this reason, many researchers are focusing on solutions to secure UAV systems against cyber-attacks and data leakage. Examples of such solutions are shown in Table 4.4.

Table 4.4 Solutions to protect UAV systems

Solutions	Publication example
Privacy-preserving schemes	[196]
Systems to monitor network traffic and detect intrusion	[197]
Identity authentication protocols	[198]
Systems for modelling attacks against cyber-physical systems	[199]

4.3.6 Preventive Activities Against the Risks Associated with Drones

As drones represent a source of risk in many situations, a significant research trend is developing technologies and solutions to counter the risks associated with drones. A critical issue in this case is the early detection and recognition of a drone, which is a complex task due to its small size and similarity to birds. For this reason, various solutions for early drone detection are analysed in publications. Some of these are presented in Table 4.5.

In addition to detection systems, other solutions for securing the correct operation of drones and enhancing their exploitation are also being investigated. Examples of such solutions described in the literature are shown in Table 4.6.

The cause of disruptions occurring during the execution of missions by drones may be the people responsible for their control. For this reason, as a preventive measure, researchers are analysing:

- Who are the operators, and what are the characteristics of commercial drone users [216]?
- The behaviours of drone operators and the risk mitigation methods they use [217].
- Screening and assessment systems for drone pilot trainees [218].
- Opportunities to use a virtual environment for drone control [219].

Table 4.5 Drone detection systems

Drone detection solutions	Publication example
Deep machine learning	[200, 201]
Convolutional deep convolutional neural networks (DC-CNNs)	[202]
Space–time information and optical flows	[203]
Radio frequencies (RF)	[204]
Sensors that measure the sound emitted by the UAV	[205]
Transformer networks	[169]
Fisheye camera systems	[206]

Table 4.6 Systems to increase the safety of drone operations

System	Publication example
Systems to detect the abnormal status of the drone	[207]
Collision avoidance systems	[208–213]
Systems to block drone access to controlled airspace such as power plants, airports, military facilities	[214]
Air traffic enhancement procedures	[215]

4.3.7 Survey Opinion

The growing popularity of drones and their application in many areas of the daily life of citizens necessitates gaining social acceptance for their functioning in a space shared with people. Consequently, research aimed at social evaluation is an important category:

- The use of drones as a delivery service [126, 220–223].
- The use of drones in rescue operations on beaches [224].
- The possible changes in users' behavioural intentions [126].
- The factors influence a change in the perception of drones [221, 225].

These studies' results may further develop UAV systems or limit their use in space shared with humans.

4.4 Limitations of the Risk Analysis Methods Used in Internal Transport in Logistics 4.0

The literature review concluded that risk aspects are an important part of ongoing research into the operation of AGV systems, as the number of such publications increases each year. However, a critical analysis of the highlighted publications shows that despite the selection criterion adopted in the form of the keyword "risk", only a few publications directly refer to the assessment or management of the risks associated with the operation of AGVs. Where studies aimed directly at risk assessment have appeared in the collection of papers analysed, they have tended to address aspects related to the safety of their movement over designated terrain. Therefore, the published results of the analyses concerned the optimisation of the paths of movement of the vehicles, their navigation and location methods, as well as the improvement of the process of coordinating their operation through optimised scheduling. However, the risk assessment methods were only a tool to analyse the benefits or risks of implementing an AGV system for the operations studied. Thus, the published research results did not address developing or improving risk assessment methods to account for their operation in complex cyber-physical systems.

Based on a comprehensive analysis of the results of the systematic literature review, the Author identified gaps in research on the risks associated with the operation of AGV systems, as shown in Fig. 4.10.

(1) **Lack of research on developing risk identification and assessment methods adapted to the specific operation of complex cyber-physical systems**

As highlighted in earlier chapters, Industry 4.0 solutions are causing not only technological but also organisational changes. The modification concerns the operating environment of anthropotechnical systems, including workspaces shared by humans and vehicles. A critical issue is the new role of digital systems, whose support affects

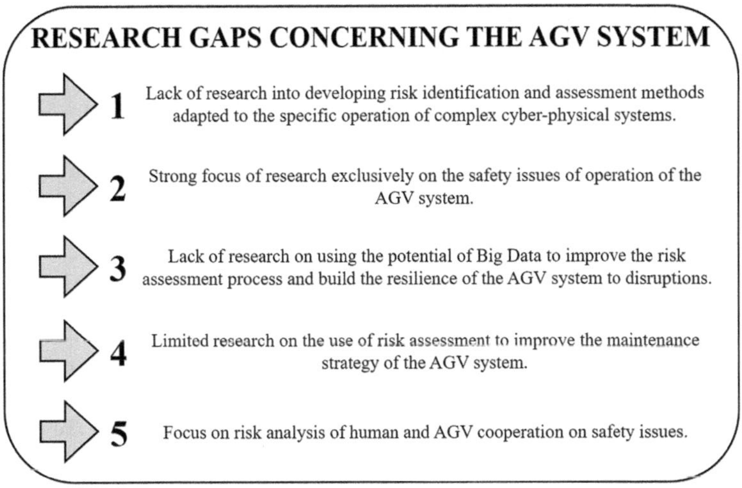

Fig. 4.10 Identified research gaps regarding risks in the operation of AGV systems

the entire working environment. The internal transport execution process is also changing, particularly regarding human-vehicle collaboration. It will force a change in the context of risk assessment, which should be strongly oriented towards identifying unknown risks that may have played a less critical role in assessment procedures. The framework of the risk assessment method to be carried out should be an interdisciplinary one, which must consider the nature of the functioning of all the elements that make up complex cyber-physical systems (i.e. people, vehicles, digital systems) and the relationships between them.

(2) **Focusing research exclusively on the safety of AGV system operation**

In the reviewed publications on risk in AGV systems, the main focus of researchers has been on adverse events related to the safety of their operation. Therefore, in research categories such as navigation, path planning, and scheduling, researchers mainly analysed the risk of possible collisions between vehicles or between vehicles and other objects (including people). The risk approach presented aligns with the engineering approach described in Sect. 2.2. A second important aspect related to research on the safety of AGV systems has been assessing the risk of cyber-attacks and their impact on the operation of the handling system. However, most of these publications have focused on the type of attack and the mechanisms to counteract its occurrence rather than a detailed analysis of its impact on process execution parameters. In this regard, it should be noted that the literature review shows a lack of a holistic view of the risk analysis carried out and that research is carried out exclusively in the engineering stream (lack of a process approach).

(3) **The lack of research on using the potential of Big Data to improve the risk assessment process and build the resilience of the AGV system**

As highlighted on several occasions, particularly in Chap. 3, Big Data structures' data collection and analysis capabilities offer tremendous opportunities regarding monitoring and multi-criteria risk assessment of the operation of IoT systems, including operational analyses as a basis for proactive equipment operation. However, these topics are covered to a limited degree in the publications analysed. Meanwhile, a critical issue in this area is research on the structures and scope of collected data adapted to the needs of monitoring and risk assessment related to the operation and the parameters of mission accomplishment by AGVs.

(4) **Limited research on using risk assessment to improve AGV system maintenance strategies**

The continuity of logistics service largely depends on the reliability of the internal transport system. Nowadays, more and more research is directed towards improving maintenance strategies based on big data and risk analyses to predict the frequency and type of damage that may occur during the operation of technical systems. The leading strategies here are Risk Based Inspection and Risk Based Maintenance (examples of the Author's research on these concepts can be found in [226]). However, in the publications analysed, no paper was found on applying one of these strategies to the operation of AGV systems. Also, studies by other authors indicate the existence of this research gap. An example is Yan's research [227], which highlights that the reliability issues and maintenance strategies of AGVs have not previously been studied sufficiently.

(5) **The focus of human-AGV collaboration risk analysis on safety issues**

As noted above, most reviewed publications on human-AGV cooperation focus on assessing the risk of vehicle-human collisions. This line of research is typical of the engineering approach, according to which the risk assessment carried out is primarily to protect human life and health. However, in the process approach, such a narrow analysis of adverse events makes it impossible to carry out a comprehensive risk assessment to build resilience in the internal transport system. The studies on human-AGV interaction highlight the risks associated with the mutual communication between man and system, particularly the issues concerning the real-time exchange of messages between man and vehicle. It is also worth noting that the adverse events occur due to human error disrupting the whole system. The AGV system relies on mission execution algorithms formulated for the correct process. Some human errors can be considered in the prepared protection system (e.g. human intrusion into the vehicle's path). However, not all errors can be predicted by the designers. It confirms the critical need for interdisciplinary teams to prepare risk analyses. The results of these analyses can be used to develop contingency solutions, which should be incorporated into procedures and systems for securing the man-against-vehicle system. There is also limited research on the risks associated with employees resisting the need to cooperate with AGV systems and taking deliberate action to disrupt the system. Although this phenomenon is marginal, such sabotaging attitudes of employees should also be considered when planning implementation activities.

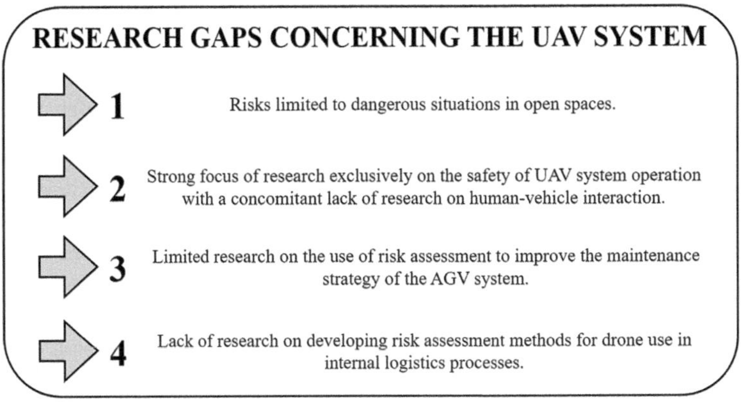

Fig. 4.11 Identified research gaps regarding risks in the operation of UAV systems

The literature review on drone risks also did not yield satisfactory results. First of all, it is worth noting that, despite a well-chosen search for publications on risk and drones simultaneously, risk as a subject of research appeared in only a few papers. At the same time, publications focused on risk assessment usually did not deal with the UAV system itself but referred to selected phenomena under observation, and drones were usually used only for data collection. The additional analysis confirmed the results of the first analysis, proving that in only 19 papers out of 257, UAV systems were the subject of study. In the remaining 238 papers, drones were only a tool for data collection, which focused on totally different research areas. The 19 identified publications on UAV systems belonged to the categories "Cyber security" and "Drones as a source of risk". The results of this analysis indicate a clear research gap, which is illustrated in detail in Fig. 4.11.

(1) **Risks limited to dangerous situations in open spaces**

The few publications that assessed the risks associated with drone operations were mainly concerned with their use in open spaces (missions carried out in the field). Papers describing drones as a source of risk investigated their impacts on bird and animal sightings and possibly dangerous situations associated with drone movement in airspace. Only isolated papers dealt with using drones in confined spaces (mainly a warehouse), but only one paper directly addressed the risk assessment of process handling (one of the Author's papers). The execution of a mission by a drone in an enclosed space (e.g. warehouse, production hall) differs from its movement in an open space. This is confirmed by the Author's research [159], which demonstrates that using drones in confined space requires the preparation of a suitable working environment. The lack of introduction of appropriate signage, provision of proper lighting and temperature, and availability of digital communications significantly affect the performance of missions performed by the UAV system. Therefore, the risk assessment for the use of drones as an IoT requires a change in the context of the assessment to be carried out and to take into account the different conditions of the working environment in which the drone carries out the mission.

(2) **Strong focus of research solely on the safety of UAV system operation, with a concomitant lack of research on human-vehicle interaction**

As in the case of the AGV system, research on the risks associated with drone operation is also dominated by an engineering approach. The best evidence of this is the highlighted category "Drones as a source of risk", which examines the safety of drone use in various environments. It is noteworthy, however, that none of the reviewed papers describe research on human-drone cooperation. The only research on drone movement in shared space concerned the performance of missions by a UAV system in an urban space with a high density of people. In contrast, no publications have reported how humans and drones share workspace. This is important because UAV systems, when moving above workers' heads, often worsen the comfort of their work due to a reduced sense of safety. In addition, drone rotors cause significant noise disturbance to the shared workspace, which can cause psycho-physical deterioration for those working close to them. Therefore, ensuring social acceptance of shared workspace is critical.

(3) **Limited research on using risk assessment to improve AGV system maintenance strategies**

Drones are often used to monitor the condition of critical infrastructure, and the data collected is used to assess the risks based on which maintenance work is planned. However, it is worth noting that there were no publications in the collected documents on using risk assessment to improve the maintenance strategy of UAV systems. Meanwhile, the safety of drone operations is significantly influenced by issues such as:

- Software updates—they reduce the occurrence of errors, improve compliance with operating system updates, and can also affect flight stability.
- Battery management—many experts consider this a critical component of UAV systems. Adequate monitoring of battery condition, checking for signs of degradation, determines the safety of the mission performed by the drones, as well as the life of the battery itself.
- Calibration of sensors—it determines the accuracy of operation and ensures reliability during flight.
- Analysis of the data generated by the UAV system during the execution of the mission (flight logs, sensor readings)—this allows the system's performance to be evaluated and potential adverse events to be identified.

These aspects should be the subject of risk assessment-oriented research, the results of which would allow better management of UAV system handling and improve the reliability of their operation.

(4) **Lack of research on developing risk assessment methods for using drones in internal logistics processes**

As noted above, the literature review results did not reveal any publications in which the research subject was the adaptation or development of a method for assessing the risks associated with the operation of UAV systems. In most publications analysed, drones were only a tool for carrying out studies (most often condition monitoring) related to risk assessment of specific phenomena. Even those studies that dealt with the assessment of the occurrence of adverse events, in which drones were the source of the risk involved, did not focus on the appropriate selection of tools and the definition of the context for the analysis being prepared. Only the Author's highlighted publication [159] dealt with adapting the risk assessment method to the specific conditions of UAV system operation in the storage space.

Thus, the critical analysis of the results obtained from the systematic literature review for both materials handling systems promoted under Industry 4.0 solutions confirms the existence of the research gap indicated in Chap. 1.

4.5 Summary

The review of Industry 4.0 solutions presented in Chap. 3 made it possible to distinguish the two most popular materials handling systems described within the Logistics 4.0 concept. These are AGV and UAV systems. AGV systems have been known and used in internal logistics for many years. However, the intensive development and growing interest in implementing AGV systems in internal logistics was only initiated by the growing popularity of Industry 4.0. UAV solutions have also been known for many years. However, their use was primarily for military operations in the initial development phase. It was not until the growing popularity of Industry 4.0 and the search for automated and autonomous solutions capable of handling material deliveries in smart factories that interest was aroused in the potential for drone use in internal logistics. The continuous development of both systems and their increasing use in transport operations is resulting in new risks and a changing level of risk for events that were once marginal to the material flows handled. However, today, this can cause critical disruptions in the continuity of supply.

For this reason, in Sect. 4.4, the Author presented a comprehensive literature review of the last five years to identify research trends on the risks associated with the operation of both systems. Section 4.1 presents the research results on the risks associated with the operation of AGV systems and their interaction with humans. The results described dealt with the twelve leading research trends defined for the risks associated with AGV systems. The results of the systematic literature analysis indicated that a critical research issue is the operational safety of AGV systems. Therefore, this is the leading research trend in the risk area concerning the operation of AGV systems. Safety aspects are also subordinated to other identified research trends, such as localisation, navigation, path planning and scheduling. Among the issues described in this topic, research on potential collision situations should be considered a critical risk aspect. This research concerns collisions between vehicles or between vehicles and other objects (including humans). For this reason, the second area of research presented in this section was the interaction of the AGV system with humans, particularly the issues of sharing the workspace. Here, three main research trends were identified concerning:

- Designing a safe work environment. This research trend focuses primarily on the safety aspects of human-vehicle interaction. Therefore, the articles in this group dealt with aspects related to human detection systems, vehicle navigation in a shared workspace and safety validation systems.
- Human-AGV cooperation. This research trend focuses primarily on ensuring efficient and safe human-AGV cooperation. Therefore, articles belonging to this group presented research results related to improving communication systems between the operator and the AGV system, considering worker behaviour and potential human error in the AGV system's operation, and optimising the shared workspace.
- Comparison of AGV and human work. This research trend is primarily concerned with identifying risks and limitations associated with human work that can be automated by implementing an AGV system and reducing the risks associated with the human factor.

A similar literature analysis was carried out for the UAV system, and the results are presented in Sect. 4.2. The results described dealt with the seven leading research trends for risks associated with UAV systems. The research on the risks associated with this system is quite different. The safety aspects of drone operations represent one of the smaller sets of papers analysed. In these studies, UAV systems are treated as a potential source of hazards, and the risk aspects focus on potentially dangerous situations involving them in urban and aerial spaces. Also related to this theme is the second of the identified research trends concerning public opinion surveys about using drones in various situations where they would have to share space with humans. The most popular research trend identified in the review on the risks associated with UAV systems is monitoring. Publications in this group present results on using drones in measurements that complement field surveys or act as a monitoring function reporting the need for field surveys. However, a critical analysis of the results described in the papers in this group proved that the purpose of the research carried out is not issues related to the risk of drone use but the observation of other phenomena and that drones only play the role of a tool for taking measurements (collecting data).

The results of the critical analysis of all three literature reviews, conducted using the PRISMA method, identified research gaps that currently exist in the area of risk assessment methods for the use of automated transport systems. These limitations for the two systems analysed are detailed in Sect. 4.3. Summarising the results obtained from the theoretical research, the Author indicates that there is currently a lack of publications on direct studies related to the selection of operational risk assessment methods, considering the new operating conditions of internal logistics processes supported by complex cyber-physical systems. A detailed literature review of two leading internal transport systems proved that:

Table 4.7 Identified research gap

Lack of research on the adaptation of operational risk assessment methods to the changes in logistics processes, in particular internal transport, caused by their handling by CHT systems, which would take into account an interdisciplinary process approach to risk analysis

- Risk assessment in publications on AGV and UAV systems is only an analytical tool used to achieve the primary research objective—there is a lack of research on adapting risk assessment methods to the new process execution conditions and specific operating environments.
- The range of risk assessments carried out for the operation of both transport solutions is very wide, but individual publications (analyses) refer to a selected fragment of the system (silo approach)—no comprehensive, interdisciplinary approach to risk assessment.
- Risk assessment usually refers to events occurring in the execution or planning phase. No studies have been identified in which the analytical proceedings include a classification of events based on the phases of the process management cycle in the risk identification and analysis phase, which can then determine how risks are managed.

In addition, it is also worth noting that the concepts of risk management in companies discussed in Chap. 2 also do not yet consider the impact of digital transformation on the changing conditions for implementing operational processes. At the same time, there is a lack of publications presenting the risk analysis results with an interdisciplinary character and combining business and engineering approaches. Meanwhile, such a combination seems natural in the case of logistics and transport processes, which are now being handled by increasingly complex cyber-physical systems.

The overall research investigation is summarised in the identified research gap shown in Table 4.7 and Fig. 4.12.

The identified research gap has set a scientific goal for further research conducted by the Author. It is developing an interdisciplinary method of operational risk assessment for logistics processes, particularly internal transport, supported by complex cyber-human-technical systems.

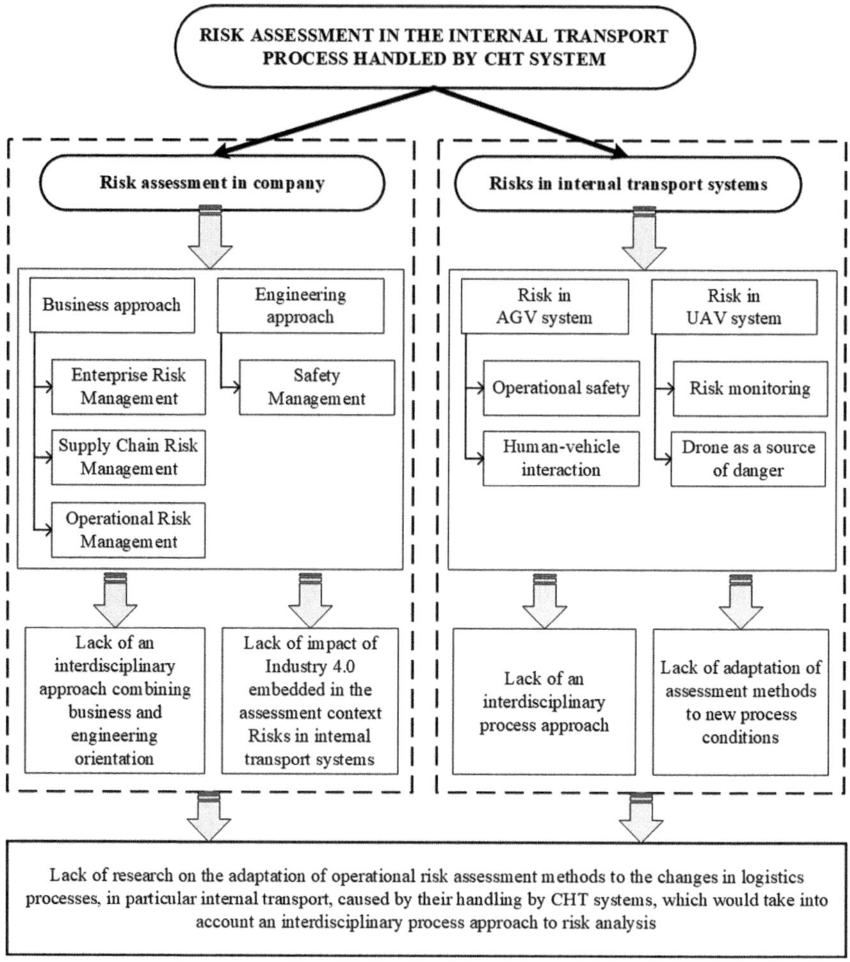

Fig. 4.12 Identified research gap

References

1. Tubis AA, Poturaj H (2022) Risk related to AGV systems—open-access literature review. Energies (Basel) 15:8910. https://doi.org/10.3390/en15238910
2. Tubis AA, Poturaj H (2021) Challenges in the implementation of autonomous robots in the process of feeding materials on the production line as part of Logistics 4.0. Logforum 17:411–423 (2021). https://doi.org/10.17270/J.LOG.2021.611
3. Tubis AA, Poturaj H, Smok A (2024) Interaction between a human and an AGV system in a shared workspace—a literature review identifying research areas. Sustainability 16:974. https://doi.org/10.3390/su16030974
4. Zhang H, Xu L, Liang J, Sun X (2022) Research on guide line identification and lateral motion control of AGV in complex environments. Machines 10:121. https://doi.org/10.3390/machines10020121

5. Sun S, Hu J, Li J, Liu R, Shu M, Yang Y (2019) An INS-UWB based collision avoidance system for AGV. Algorithms 12:40. https://doi.org/10.3390/a12020040
6. Adriaensen A, Pintelon L, Costantino F, Di Gravio G, Patriarca R (2021) An STPA safety analysis case study of a collaborative robot application. IFAC-PapersOnLine 54:534–539. https://doi.org/10.1016/J.IFACOL.2021.08.061
7. Zamora-Cadenas L, Velez I, Sierra-Garcia JE (2021) UWB-based safety system for autonomous guided vehicles without hardware on the infrastructure. IEEE Access 9:96430–96443. https://doi.org/10.1109/ACCESS.2021.3094279
8. Zhang Z, Guo Q, Chen J, Yuan P (2018) Collision-free route planning for multiple AGVs in an automated warehouse based on collision classification. IEEE Access 6:26022–26035. https://doi.org/10.1109/ACCESS.2018.2819199
9. Klimenda F, Cizek R, Pisarik M, Sterba J (2021) Stopping the mobile robotic vehicle at a defined distance from the obstacle by means of an infrared distance sensor. Sensors 21:5959. https://doi.org/10.3390/s21175959
10. Kutzler T, Wolter A, Kenner A, Dassow S (2021) Boosting cyber-physical system security. IFAC-PapersOnLine 54:976–981. https://doi.org/10.1016/J.IFACOL.2021.08.117
11. Elsisi M, Tran M-Q (2021) Development of an IoT architecture based on a deep neural network against cyber attacks for automated guided vehicles. Sensors 21:8467. https://doi.org/10.3390/s21248467
12. Garnier S, Subrin K, Arevalo-Siles P, Caverot G, Furet B (2018) Mobile robot stability for complex tasks in naval industries. Procedia CIRP 72:297–302. https://doi.org/10.1016/j.procir.2018.03.101
13. Zhang H, Qi L, Luan W, Ma H (2022) Double-cycling AGV scheduling considering uncertain crane operational time at container terminals. Appl Sci 12:4820. https://doi.org/10.3390/app12104820
14. Guo K, Zhu J, Shen L (2021) An Improved acceleration method based on multi-agent system for AGVs conflict-free path planning in automated terminals. IEEE Access 9:3326–3338. https://doi.org/10.1109/ACCESS.2020.3047916
15. Yuan Z, Yang Z, Lv L, Shi Y (2020) A bi-level path planning algorithm for multi-AGV routing problem. Electronics (Basel) 9:1351. https://doi.org/10.3390/electronics9091351
16. Wang J, Luo Y, Tan X (2021) Path planning for automatic guided vehicles (AGVs) fusing MH-RRT with improved TEB. Actuators 10:314. https://doi.org/10.3390/act10120314
17. Meysami A, Cuillière J-C, François V, Kelouwani S (2022) Investigating the impact of triangle and quadrangle mesh representations on AGV path planning for various indoor environments: with or without inflation. Robotics 11:50. https://doi.org/10.3390/robotics11020050
18. Szczepanski R, Tarczewski T, Erwinski K (2022) Energy efficient local path planning algorithm based on predictive artificial potential field. IEEE Access 10:39729–39742. https://doi.org/10.1109/ACCESS.2022.3166632
19. Wu B, Chi X, Zhao C, Zhang W, Lu Y, Jiang D (2022) Dynamic path planning for Forklift AGV based on smoothing A* and improved DWA hybrid algorithm. Sensors 22:7079. https://doi.org/10.3390/s22187079
20. Run R-S, Xiao Z-Y (2018) Indoor autonomous vehicle navigation—a feasibility study based on infrared technology. Appl Syst Inno 1:4. https://doi.org/10.3390/asi1010004
21. Cabezas-Olivenza M, Zulueta E, Sánchez-Chica A, Teso-Fz-Betoño A, Fernandez-Gamiz U (2021) Dynamical analysis of a navigation algorithm. Mathematics 9:3139. https://doi.org/10.3390/math9233139
22. Reger M, Stumpenhausen J, Bernhardt H (2022) Evaluation of LiDAR for the free navigation in agriculture. AgriEngineering 4:489–506. https://doi.org/10.3390/agriengineering4020033
23. Nguyen PT-T, Yan S-W, Liao J-F, Kuo C-H (2021) Autonomous mobile robot navigation in sparse LiDAR feature environments. Appl Sci 11:5963. https://doi.org/10.3390/app11135963
24. Hu X, Luo Z, Jiang W (2020) AGV localization system based on ultra-wideband and vision guidance. Electronics (Basel) 9:448. https://doi.org/10.3390/electronics9030448
25. An X, Zhao S, Cui X, Shi Q, Lu M (2020) Distributed multi-antenna positioning for automatic-guided vehicle. Sensors 20:1155. https://doi.org/10.3390/s20041155

26. De Ryck M, Versteyhe M, Debrouwere F (2020) Automated guided vehicle systems, state-of-the-art control algorithms and techniques. J Manuf Syst 54:152–173. https://doi.org/10.1016/J.JMSY.2019.12.002
27. De Ryck M, Pissoort D, Holvoet T, Demeester E (2021) Decentral task allocation for industrial AGV-systems with resource constraints. J Manuf Syst 59:310–319. https://doi.org/10.1016/J.JMSY.2021.03.008
28. Barak S, Moghdani R, Maghsoudlou H (2021) Energy-efficient multi-objective flexible manufacturing scheduling. J Clean Prod 283:124610. https://doi.org/10.1016/J.JCLEPRO.2020.124610
29. Tang G, Tang C, Claramunt C, Hu X, Zhou P (2021) Geometric A-star algorithm: an improved A-star algorithm for AGV path planning in a port environment. IEEE Access 9:59196–59210. https://doi.org/10.1109/ACCESS.2021.3070054
30. Han W, Xu J, Sun Z, Liu B, Zhang K, Zhang Z, Mei X (2022) Digital Twin-based automated guided vehicle scheduling: a solution for its charging problems. Appl Sci 12:3354. https://doi.org/10.3390/app12073354
31. Witczak M, Lipiec B, Banaszak Z (2019) Fault-tolerant control-based flexible AGV transportation in a seat assembly system. IFAC-PapersOnLine 52:67–72. https://doi.org/10.1016/j.ifacol.2019.10.028
32. Xu R, Feng H, Liu J, Hong W (2022) Dynamic spare point application based coordination strategy for multi-AGV systems in a WIP warehouse environment. IEEE Access 10:80249–80263. https://doi.org/10.1109/ACCESS.2022.3195173
33. Lyu X, Song Y, He C, Lei Q, Guo W (2019) Approach to integrated scheduling problems considering optimal number of automated guided vehicles and conflict-free routing in flexible manufacturing systems. IEEE Access 7:74909–74924. https://doi.org/10.1109/ACCESS.2019.2919109
34. Mrugalska B, Stetter R (2019) Health-aware model-predictive control of a cooperative AGV-based production system. Sensors 19:532. https://doi.org/10.3390/s19030532
35. Dang QV, Singh N, Adan I, Martagan T, van de Sande D (2021) Scheduling heterogeneous multi-load AGVs with battery constraints. Comput Oper Res 136:105517. https://doi.org/10.1016/J.COR.2021.105517
36. Bohács G, Győrváry Zs, Gáspár D (2021) Integrating scheduling and energy efficiency aspects in production logistic using AGV systems. IFAC-PapersOnLine 54:294–299. https://doi.org/10.1016/j.ifacol.2021.08.156
37. Yue L, Fan H, Zhai C (2019) Joint configuration and scheduling optimization of a dual-trolley quay crane and automatic guided vehicles with consideration of vessel stability. Sustainability 12:24. https://doi.org/10.3390/su12010024
38. Sperling M, Kivelä T (2022) Concept of a dual energy storage system for sustainable energy supply of automated guided vehicles. Energies (Basel) 15:479. https://doi.org/10.3390/en15020479
39. Niestrój R, Rogala T, Skarka W (2020) An energy consumption model for designing an AGV energy storage system with a PEMFC stack. Energies (Basel) 13:3435. https://doi.org/10.3390/en13133435
40. Hanschek AJ, Bouvier YE, Jesacher E, Grbović PJ (2022) Analysis and comparison of power distribution system topologies for low-voltage DC–DC automated guided vehicle applications. Energies (Basel) 15, 2012. https://doi.org/10.3390/en15062012
41. Chen W, Liu J, Chen S, Zhang L (2020) Energy shaping control for wireless power transfer system in automatic guided vehicles. Energies (Basel) 13:2959. https://doi.org/10.3390/en13112959
42. Liang C, Zhang Y, Li Z, Yuan F, Yang G, Song K (2020) Coil positioning for wireless power transfer system of automatic guided vehicle based on magnetic sensing. Sensors 20:5304. https://doi.org/10.3390/s20185304
43. Lu F, Zhang Y, Zhang H, Zhu C, Diao L, Gong M, Zhang W, Mi C (2019) A low-voltage and high-current inductive power transfer system with low harmonics for automatic guided vehicles. IEEE Trans Veh Technol 68:3351–3360. https://doi.org/10.1109/TVT.2019.2898418

44. Fellan A, Schellenberger C, Zimmermann M, Schotten HD (2018) Enabling communication technologies for automated unmanned vehicles in Industry 4.0. In: 2018 International conference on information and communication technology convergence (ICTC). IEEE, pp 171–176. https://doi.org/10.1109/ICTC.2018.8539695
45. Nakimuli W, Garcia-Reinoso J, Sierra-Garcia JE, Serrano P, Fernandez IQ (2021) Deployment and evaluation of an Industry 4.0 use case over 5G. IEEE Commun Mag 59:14–20. https://doi.org/10.1109/MCOM.001.2001104
46. Cheong HW, Lee H (2018) Requirements of AGV (automated guided vehicle) for SMEs (small and medium-sized enterprises). Procedia Comput Sci 139:91–94. https://doi.org/10.1016/J.PROCS.2018.10.222
47. Seder M, Petrović L, Peršić J, Popović G, Petković T, Šelek A, Bićanić B, Cvišić I, Josić D, Marković I, Petrović I, Muhammad A (2019) Open platform based mobile robot control for automation in manufacturing logistics. IFAC-PapersOnLine 52:95–100. https://doi.org/10.1016/j.ifacol.2019.11.055
48. Martínez-Gutiérrez A, Díez-González J, Ferrero-Guillén R, Verde P, Álvarez R, Perez H (2021) Digital Twin for automatic transportation in Industry 4.0. Sensors 21:3344. https://doi.org/10.3390/s21103344
49. Azangoo M, Taherkordi A, Blech JO, Vyatkin V (2021) Digital Twin-assisted controlling of AGVs in flexible manufacturing environments. In: 2021 IEEE 30th International symposium on industrial electronics (ISIE). IEEE, pp 1–7. https://doi.org/10.1109/ISIE45552.2021.9576361
50. Alexandru M, Dragos C, Bala-Constantin Z (2022) Digital Twin for automated guided vehicles fleet management. Procedia Comput Sci 199:1363–1369. https://doi.org/10.1016/J.PROCS.2022.01.172
51. Allmacher C, Schumann M, Klimant P, Putz M (2019) Optimizing development time through hybrid commissioning of control software. Procedia CIRP 79:450–455. https://doi.org/10.1016/J.PROCIR.2019.02.120
52. Kłosowski G, Gola A, Amila T (2018) Computational intelligence in control of AGV multimodal systems. IFAC-PapersOnLine 51:1421–1427. https://doi.org/10.1016/J.IFACOL.2018.08.315
53. Stetter R (2020) A fuzzy virtual actuator for automated guided vehicles. Sensors 20:4154. https://doi.org/10.3390/s20154154
54. Jiang W, Zhu G, Zheng Y (2021) Iterative learning control for AGV drive motor based on linear extended state observer. Machines 9:324. https://doi.org/10.3390/machines9120324
55. Wang X, Shi H, Zhang C (2020) Path planning for intelligent parking system based on improved ant colony optimization. IEEE Access 8:65267–65273. https://doi.org/10.1109/ACCESS.2020.2984802
56. Thomopoulos V, Bitas D, Papastavros K-N, Tsipianitis D, Kavga A (2021) Development of an integrated IoT-based greenhouse control three-device robotic system. Agronomy 11:405. https://doi.org/10.3390/agronomy11020405
57. Park S, Hwang J, Yang H, Kim S (2021) Simulation modelling for automated guided vehicle introduction to the loading process of ro-ro ships. J Mar Sci Eng 9:441. https://doi.org/10.3390/jmse9040441
58. Kim J (2022) Fast route planner considering terrain information. Sensors 22:4518. https://doi.org/10.3390/s22124518
59. Zaccaria M, Giorgini M, Monica R, Aleotti J (2021) Multi-robot multiple camera people detection and tracking in automated warehouses. In: 2021 IEEE 19th International conference on industrial informatics (INDIN). IEEE, pp 1–6. https://doi.org/10.1109/INDIN45523.2021.9557363
60. Indri M, Sibona F, David Cen Cheng P (2020) Sen3Bot Net: a meta-sensors network to enable smart factories implementation. In: 2020 25th IEEE international conference on emerging technologies and factory automation (ETFA). IEEE, pp 719–726. https://doi.org/10.1109/ETFA46521.2020.9212125

61. Ahmed HO (2020) FLS-based collision avoidance cyber physical system for warehouse robots using FPGA. In: 2019 6th International conference on dependable systems and their applications (DSA). IEEE, pp 262–268. https://doi.org/10.1109/DSA.2019.00040
62. Rey R, Corzetto M, Cobano JA, Merino L, Caballero F (2019) Human-robot co-working system for warehouse automation. In: 2019 24th IEEE international conference on emerging technologies and factory automation (ETFA). IEEE, pp 578–585. https://doi.org/10.1109/ETFA.2019.8869178
63. Écorchard G, Košnar K, Přeučil L (2020) Wearable camera-based human absolute localization in large warehouses. In: Osten W, Nikolaev DP (eds) Twelfth international conference on machine vision (ICMV 2019). SPIE, p 96. https://doi.org/10.1117/12.2559424
64. Holzke F, Danielis P, Golatowski F, Timmermann D (2018) A fusion approach for the localization of humans in factory environments. In: 2018 IEEE industrial cyber-physical systems (ICPS). IEEE, pp 59–64. https://doi.org/10.1109/ICPHYS.2018.8387638
65. Babic J, Bilic M, Kovac I (2022) Safety vest system for human-robot collaboration. In: 2022 45th Jubilee international convention on information, communication and electronic technology (MIPRO). IEEE, pp 12–17. https://doi.org/10.23919/MIPRO55190.2022.9803318
66. Sabattini L, Aikio M, Beinschob P, Boehning M, Cardarelli E, Digani V, Krengel A, Magnani M, Mandici S, Oleari F, Reinke C, Ronzoni D, Stimming C, Varga R, Vatavu A, Castells Lopez S, Fantuzzi C, Mayra A, Nedevschi S, Secchi C, Fuerstenberg K (2018) The PAN-robots project: advanced automated guided vehicle systems for industrial logistics. IEEE Robot Autom Mag 25:55–64. https://doi.org/10.1109/MRA.2017.2700325
67. Lambert ED, Romano R, Watling D (2021) Optimal smooth paths based on clothoids for car-like vehicles in the presence of obstacles. Int J Control Autom Syst 19:2163–2182. https://doi.org/10.1007/s12555-020-0179-1
68. Sun T, Wang Z, Li Q (2019) A medical garbage bin recycling system based on AGV. In: Proceedings of the 2019 international conference on robotics, intelligent control and artificial intelligence—RICAI 2019. ACM Press, New York, New York, USA, pp 220–225. https://doi.org/10.1145/3366194.3366232
69. Millán M, Sierra-García JE, Santos M (2022) Generation of restricted zones for AGVs routes by clustering algorithms. https://doi.org/10.1007/978-3-030-87869-6_45
70. Elgeziry M, Costa F, Genovesi S (2022) Radio-frequency guidance system for path-following industrial autonomous guided vehicles. In: 2022 16th European conference on antennas and propagation (EuCAP). IEEE, pp 1–5. https://doi.org/10.23919/EuCAP53622.2022.9769288
71. Tien TN, Nguyen K-V (2022) Updated weight graph for dynamic path planning of multi-AGVs in healthcare environments. In: 2022 International conference on advanced technologies for communications (ATC). IEEE, pp 130–135. https://doi.org/10.1109/ATC55345.2022.9943032
72. Papcun P, Cabadaj J, Kajati E, Romero D, Landryova L, Vascak J, Zolotova I (2019) Augmented reality for humans-robots interaction in dynamic slotting "chaotic storage" smart warehouses. https://doi.org/10.1007/978-3-030-30000-5_77
73. Plosz S, Varga P (2018) Security and safety risk analysis of vision guided autonomous vehicles. In: 2018 IEEE industrial cyber-physical systems (ICPS). IEEE, pp 193–198. https://doi.org/10.1109/ICPHYS.2018.8387658
74. Anastasi S, Monica L, Madonna M, Di Nardo M (2021) Human—collaborative machine interaction: the effects on the standardization. In: Proceedings of the 31st European safety and reliability conference (ESREL 2021). Research Publishing Services, Singapore, pp 3406–3410. https://doi.org/10.3850/978-981-18-2016-8_415-cd
75. Fiolka M, Jost J, Kirks T (2020) Explorative study of a brain-computer interface for order picking tasks in logistics. In: 2020 12th International conference on intelligent human-machine systems and cybernetics (IHMSC). IEEE, pp 253–256. https://doi.org/10.1109/IHMSC49165.2020.10135
76. Wang Y-X, Chang C-L (2020) ROS-base multi-sensor fusion for accuracy positioning and SLAM system. In: 2020 International symposium on community-centric systems (CcS). IEEE, pp 1–6. https://doi.org/10.1109/CcS49175.2020.9231442

77. Gong X, Wang T, Huang T, Cui Y (2022) Toward safe and efficient human swarm collaboration: a hierarchical multi-agent pickup and delivery framework. IEEE Trans Intell Veh:1–1. https://doi.org/10.1109/TIV.2022.3172342
78. Malayjerdi M, Kuts V, Sell R, Otto T, Baykara BC (2020) Virtual simulations environment development for autonomous vehicles interaction. In: Volume 2B: Advanced manufacturing. American Society of Mechanical Engineers. https://doi.org/10.1115/IMECE2020-23362
79. Nguyen TP, Nguyen H, Thinh Ngo HQ (2022) Planning the emergency collision avoidance strategy based on personal zones for safe human-machine interaction in smart cyber-physical system. Complexity 2022:1–21. https://doi.org/10.1155/2022/2992379
80. Soltani A, Stonis M, Overmeyer L (2019) Development of a case-based reasoning expert system for the disturbance management in automated guided vehicle systems. Logistics J 2019. https://doi.org/10.2195/lj_Proc_soltani_en_201912_01
81. Bergman M, Bedaf S, van Heel G, Sturm J (2020) Can I just pass by? Testing design principles for industrial transport robots. In: Proceedings of the 4th international conference on computer-human interaction research and applications. SCITEPRESS—Science and Technology Publications, pp 178–187. https://doi.org/10.5220/0010144301780187
82. Löcklin A, Dettinger F, Artelt M, Jazdi N, Weyrich M (2022) Trajectory prediction of workers to improve AGV and AMR operation based on the manufacturing schedule. Procedia CIRP 107:283–288. https://doi.org/10.1016/j.procir.2022.04.046
83. Bach S-H, Yi S-Y (2022) An efficient approach for line-following automated guided vehicles based on fuzzy inference mechanism. J Robot Control (JRC) 3:395–401. https://doi.org/10.18196/jrc.v3i4.14787
84. Prati E, Peruzzini M, Pellicciari M, Raffaeli R (2021) How to include user eXperience in the design of human-robot interaction. Robot Comput Integr Manuf. 68:102072. https://doi.org/10.1016/j.rcim.2020.102072
85. Vlachos I, Pascazzi RM, Ntotis M, Spanaki K, Despoudi S, Repoussis P (2022) Smart and flexible manufacturing systems using autonomous guided vehicles (AGVs) and the Internet of Things (IoT). Int J Prod Res:1–22. https://doi.org/10.1080/00207543.2022.2136282
86. David A, Birtel M, Wagner A, Ruskowski M (2020) Architecture concept for the integration of cyber-physical transport modules in modular production environments. Procedia Manuf 51:1111–1116. https://doi.org/10.1016/j.promfg.2020.10.156
87. Mohsin S, Khan IA, Ali M (2019) Ergonomics-based working flexibility for automated guided vehicle (AGV) operators. Int J Adv Manuf Technol 103:529–547. https://doi.org/10.1007/s00170-019-03542-3
88. Ballal S, Jagannath M, Arun Venkatesh K (2019) Wireless data acquisition and communication system for automated guided vehicle. https://doi.org/10.1007/978-981-10-8663-2_25
89. Luchetti A, Carollo A, Santoro L, Nardello M, Brunelli D, Bosetti P, De Cecco M (2021) Human identification and tracking using ultra-wideband-vision data fusion in unstructured environments. ACTA IMEKO 10:124. https://doi.org/10.21014/acta_imeko.v10i4.1139
90. Allmacher C, Dudczig M, Knopp S, Klimant P (2019) Virtual reality for virtual commissioning of automated guided vehicles. In: 2019 IEEE conference on virtual reality and 3D user interfaces (VR). IEEE, pp 838–839. https://doi.org/10.1109/VR.2019.8797981
91. Yu Y-S, Yu P-Y, Chen C-W (2020) An indoor positioning technique for on-demand AGV calling system. In: 2020 IEEE international conference on consumer electronics—Taiwan (ICCE-Taiwan). IEEE, pp 1–2. https://doi.org/10.1109/ICCE-Taiwan49838.2020.9258150
92. Lee S-J, Kim W, Lee YK, Yoon D, Wook LJ, Persico G, Fischer A (2019) Design of the operator tracing robot for material handling. In: 2019 International conference on information and communication technology convergence (ICTC). IEEE, pp 1254–1256. https://doi.org/10.1109/ICTC46691.2019.8939957
93. Lai YL, Lai YK, Lan LC, Zheng CY, Chen SC, Tseng LW (2021) A novel automated guided vehicle for guidance applications. J Phys Conf Ser 2020:012037. https://doi.org/10.1088/1742-6596/2020/1/012037
94. Ding I-J, Juang Y-C (2022) Hand-gesture-control-based navigation using wearable armband with surface electromyography and inertial measurement unit sensor data for autonomous

guided vehicles with robot operation system-based simultaneous localization and mapping navigation in smart manufacturing. Sens Mater 34:3513. https://doi.org/10.18494/SAM4045
95. Manitsaris S, Senteri G, Makrygiannis D, Glushkova A (2020) Human movement representation on multivariate time series for recognition of professional gestures and forecasting their trajectories. Front Robot AI 7. https://doi.org/10.3389/frobt.2020.00080
96. Coelho F, Macedo R, Relvas S, Barbosa-Póvoa A (2022) Simulation of in-house logistics operations for manufacturing. Int J Comput Integr Manuf 35:989–1009. https://doi.org/10.1080/0951192X.2022.2027521
97. Yokota T (2019) Min-max-strategy-based optimum co-operative picking with AGVs in warehouse. In: 2019 58th Annual conference of the society of instrument and control engineers of Japan (SICE). IEEE, pp 236–242. https://doi.org/10.23919/SICE.2019.8859959
98. Zou Y, Zhang D, Qi M (2019) Order picking system optimization based on picker-robot collaboration. In: Proceedings of the 2019 5th international conference on industrial and business engineering. ACM, New York, NY, USA, pp 1–6. https://doi.org/10.1145/3364335.3364386
99. Löffler M, Boysen N, Schneider M (2022) Picker routing in AGV-assisted order picking systems. Informs J Comput 34:440–462. https://doi.org/10.1287/ijoc.2021.1060
100. Park SH, Hwang J, Yun S, Kim S (2022) Automatic guided vehicles introduction impacts to roll-on/roll-off terminals: simulation and cost model analysis. J Adv Transp 2022:1–14. https://doi.org/10.1155/2022/6062840
101. Vijayakumar V, Sgarbossa F (2021) A literature review on the level of automation in picker-to-parts order picking system: research opportunities. In: IFAC-PapersOnLine, pp 438–443. Elsevier B.V. https://doi.org/10.1016/j.ifacol.2021.08.050
102. Punekar PA, Parvati VK, Mattikalli AB, Gadad GM (2022) An innovative humanoid assistant for performing simple and repetitive tasks. In: 13th International conference on advances in computing, control, and telecommunication technologies, ACT 2022, pp 423–429
103. Ghiyasinasab M, Lahrichi N, Lehoux N (2021) A simulation model to analyse automation scenarios in decontamination centers. Health Syst:1–17. https://doi.org/10.1080/20476965.2021.2004933
104. Shimmura T, Ichikari R, Okuma T (2020) Human–Robot hybrid service system introduction for enhancing labor and robot productivity. https://doi.org/10.1007/978-3-030-57997-5_76
105. Hellmann W, Marino D, Megahed M, Suggs M, Borowski J, Negahban A (2019) Human, AGV or AIV? An integrated framework for material handling system selection with real-world application in an injection molding facility. Int J Adv Manuf Technol 101:815–824. https://doi.org/10.1007/s00170-018-2958-x
106. Shejwal Y, Behare M (2020) AGV based stretcher. In: 2020 International conference on smart innovations in design, environment, management, planning and computing (ICSIDEMPC). IEEE, pp 200–201. https://doi.org/10.1109/ICSIDEMPC49020.2020.9299646
107. Labade S (2018) Forklift free operation: compensating downtime from bullwhip effect and escalating operation safety. Presented at the April 3 (2018). https://doi.org/10.4271/2018-01-1389
108. Naeem D, Gheith M, Eltawil A (2021) Integrated scheduling of AGVs and yard cranes in automated container terminals. In: 2021 IEEE 8th international conference on industrial engineering and applications (ICIEA). IEEE, pp 632–636. https://doi.org/10.1109/ICIEA52957.2021.9436808
109. Fay B, Ramasubramanian AK, Murphy RD, Adderley T, Papakostas N (2022) Using a process simulation platform for reviewing automated airport baggage handling system configurations. Procedia CIRP 112:180–185. https://doi.org/10.1016/j.procir.2022.09.069
110. Dung BTK, Le Anh TM, Chuong TT (2021) Application of IE techniques to support businesses to make the adoption of sustainable manufacturing solutions: a case study of adoption of AGV Technology in a Garment factory. https://doi.org/10.1007/978-3-030-69610-8_78
111. Afonso T, Alves CA, Carneiro P, Vieira A (2021) Simulation pulled by the need to reduce wastes and human effort in an intralogistics project. Int J Indus Eng Manage 12:274–285. https://doi.org/10.24867/IJIEM-2021-4-294

112. Prabhakar M, Paulraj V, Dhanraj JA, Nagarajan S, Kannappan DAK, Hariharan A (2020) Design and simulation of an automated guided vehicle through Webots for isolated COVID-19 patients in hospitals. In: 2020 IEEE 4th conference on information & communication technology (CICT). IEEE, pp 1–5. https://doi.org/10.1109/CICT51604.2020.9312063
113. Zulkiflee HB, Elsheikh EMA (2022) Development sanitizer sprinkler for automated guided vehicle (AGV). In: 2022 IEEE 8th international conference on smart instrumentation, measurement and applications (ICSIMA). IEEE, pp 190–195. https://doi.org/10.1109/ICSIMA55652.2022.9928865
114. Caccavale R, Finzi A (2019) An automated guided vehicle for flexible and interactive task execution in hospital scenarios. AIRO@AI*IA
115. Tubis AA, Poturaj H, Dereń K, Żurek A (2024) Risks of drone use in light of literature studies. Sensors 24:1205. https://doi.org/10.3390/s24041205
116. Hernández-López D, López-Rebollo J, Moreno MA, Gonzalez-Aguilera D (2023) Automatic processing for identification of forest fire risk areas along high-voltage power lines using coarse-to-fine LiDAR data. Forests 14:662. https://doi.org/10.3390/f14040662
117. Munawar HS, Gharineiat Z, Akram J, Imran Khan S (2022) A framework for burnt area mapping and evacuation problem using aerial imagery analysis. Fire 5:122. https://doi.org/10.3390/fire5040122
118. Shamsoshoara A, Afghah F, Razi A, Zheng L, Fulé PZ, Blasch E (2021) Aerial imagery pile burn detection using deep learning: the FLAME dataset. Comput Netw 193:108001. https://doi.org/10.1016/j.comnet.2021.108001
119. Alsumayt A, El-Haggar N, Amouri L, Alfawaer ZM, Aljameel SS (2023) Smart flood detection with AI and blockchain integration in Saudi Arabia using drones. Sensors 23:5148. https://doi.org/10.3390/s23115148
120. Annis A, Nardi F, Petroselli A, Apollonio C, Arcangeletti E, Tauro F, Belli C, Bianconi R, Grimaldi S (2020) UAV-DEMs for small-scale flood hazard mapping. Water (Basel) 12:1717. https://doi.org/10.3390/w12061717
121. Belcore E, Piras M, Pezzoli A (2022) Land Cover Classification from Very High-Resolution UAS Data for Flood Risk Mapping. Sensors 22:5622. https://doi.org/10.3390/s22155622
122. Sodnik J, Mikoš M, Bezak N (2023) Torrential hazards' mitigation measures in a typical alpine catchment in Slovenia. Appl Sci 13:11136. https://doi.org/10.3390/app132011136
123. Tiepolo M, Belcore E, Braccio S, Issa S, Massazza G, Rosso M, Tarchiani V (2021) Method for fluvial and pluvial flood risk assessment in rural settlements. MethodsX 8:101463. https://doi.org/10.1016/j.mex.2021.101463
124. Whitehurst D, Friedman B, Kochersberger K, Sridhar V, Weeks J (2021) Drone-based community assessment, planning, and disaster risk management for sustainable development. Remote Sens (Basel) 13:1739. https://doi.org/10.3390/rs13091739
125. Jacobsen RH, Matlekovic L, Shi L, Malle N, Ayoub N, Hageman K, Hansen S, Nyboe FF, Ebeid E (2023) Design of an autonomous cooperative drone swarm for inspections of safety critical infrastructure. Appl Sci 13:1256. https://doi.org/10.3390/app13031256
126. Kim JJ, Kim I, Hwang J (2021) A change of perceived innovativeness for contactless food delivery services using drones after the outbreak of COVID-19. Int J Hosp Manag 93:102758. https://doi.org/10.1016/j.ijhm.2020.102758
127. Manzoor A, Kim K, Pandey SR, Kazmi SMA, Tran NH, Saad W, Hong CS (2021) Ruin theory for energy-efficient resource allocation in UAV-assisted cellular networks. IEEE Trans Commun 69:3943–3956. https://doi.org/10.1109/TCOMM.2021.3064968
128. Forkan ARM, Kang Y-B, Jayaraman PP, Liao K, Kaul R, Morgan G, Ranjan R, Sinha S (2022) CorrDetector: a framework for structural corrosion detection from drone images using ensemble deep learning. Expert Syst Appl 193:116461. https://doi.org/10.1016/j.eswa.2021.116461
129. Kabbabe Poleo K, Crowther WJ, Barnes M (2021) Estimating the impact of drone-based inspection on the levelised cost of electricity for offshore wind farms. Results Eng 9:100201. https://doi.org/10.1016/j.rineng.2021.100201

130. Mandirola M, Casarotti C, Peloso S, Lanese I, Brunesi E, Senaldi I (2022) Use of UAS for damage inspection and assessment of bridge infrastructures. Int J Disast Risk Reduc 72:102824. https://doi.org/10.1016/j.ijdrr.2022.102824
131. Outay F, Mengash HA, Adnan M (2020) Applications of unmanned aerial vehicle (UAV) in road safety, traffic and highway infrastructure management: recent advances and challenges. Transp Res Part A Policy Pract 141:116–129. https://doi.org/10.1016/j.tra.2020.09.018
132. Kolster ME, Wigh MD, Lima Simões da Silva E, Bjerg Vilhelmsen T, Døssing A (2022) High-speed magnetic surveying for unexploded ordnance using UAV systems. Remote Sens (Basel) 14:1134. https://doi.org/10.3390/rs14051134
133. Yoo L-S, Lee J-H, Lee Y-K, Jung S-K, Choi Y (2021) Application of a drone magnetometer system to military mine detection in the demilitarized zone. Sensors 21:3175. https://doi.org/10.3390/s21093175
134. Bridgelall R (2022) Reducing risks by transporting dangerous cargo in drones. Sustainability 14:13044. https://doi.org/10.3390/su142013044
135. Sookram N, Ramsewak D, Singh S (2021) The conceptualization of an unmanned aerial system (UAS) ship-shore delivery service for the maritime industry of Trinidad. Drones 5:76. https://doi.org/10.3390/drones5030076
136. Umeda S, Yoshikawa N, Seo Y (2022) Cost and workload assessment of agricultural drone sprayer: a case study of rice production in Japan. Sustainability 14:10850. https://doi.org/10.3390/su141710850
137. Heiets I, Kuo Y-W, La J, Yeun RCK, Verhagen W (2023) Future trends in UAV applications in the Australian market. Aerospace 10:555. https://doi.org/10.3390/aerospace10060555
138. Dams B, Chen B, Shepherd P, Ball RJ (2023) Development of cementitious mortars for aerial additive manufacturing. Appl Sci 13:641. https://doi.org/10.3390/app13010641
139. Tubis AA, Żurek AM (2022) Analysis of adverse events related to the use of drones in material handling at a chemical manufacturer site. In: Proceedings of the 32nd European safety and reliability conference. Research Publishing Services, Singapore, pp 815–821. https://doi.org/10.3850/978-981-18-5183-4_R16-03-600-cd
140. Tubis AA, Żurek A (2022) The use of drones to handle internal logistic flows in the company from the chemical sector. In: Conference proceedings—CLC 2022, June 15–17, pp 107–112. Bojnice. https://doi.org/10.37904/clc.2022.4572
141. De Silvestri S, Pagliarani M, Tomasello F, Trojaniello D, Sanna A (2022) Design of a service for hospital internal transport of urgent pharmaceuticals via drones. Drones 6:70. https://doi.org/10.3390/drones6030070
142. Barbeau M, Garcia-Alfaro J, Kranakis E (2021) Risky zone avoidance strategies for drones. In: 2021 IEEE Canadian conference on electrical and computer engineering (CCECE). IEEE, pp 1–6. https://doi.org/10.1109/CCECE53047.2021.9569172
143. Niglio F, Comite P, Cannas A, Pirri A, Tortora G (2022) Preliminary clinical validation of a drone-based delivery system in urban scenarios using a smart capsule for blood. Drones 6:195. https://doi.org/10.3390/drones6080195
144. Beck S, Bui T, Davies A, Courtney P, Brown A, Geudens J, Royall P (2020) An evaluation of the drone delivery of adrenaline auto-injectors for anaphylaxis: pharmacists' perceptions, acceptance, and concerns. Drones 4:66. https://doi.org/10.3390/drones4040066
145. Jones RW, Despotou G (2019) Unmanned aerial systems and healthcare: possibilities and challenges. In: 2019 14th IEEE conference on industrial electronics and applications (ICIEA). IEEE, pp 189–194. https://doi.org/10.1109/ICIEA.2019.8834142
146. Alsamhi SH, Shvetsov AV, Kumar S, Shvetsova SV, Alhartomi MA, Hawbani A, Rajput NS, Srivastava S, Saif A, Nyangaresi VO (2022) UAV computing-assisted search and rescue mission framework for disaster and harsh environment mitigation. Drones 6:154. https://doi.org/10.3390/drones6070154
147. Salmoral G, Rivas Casado M, Muthusamy M, Butler D, Menon P, Leinster P (2020) Guidelines for the use of unmanned aerial systems in flood emergency response. Water (Basel) 12:521. https://doi.org/10.3390/w12020521

148. Burke, McWhirter, Veitch-Michaelis, McAree, Pointon, Wich, Longmore (2019) Requirements and limitations of thermal drones for effective search and rescue in marine and coastal areas. Drones 3:78. https://doi.org/10.3390/drones3040078
149. Calamoneri T, Coro F, Mancini S (2022) A realistic model to support rescue operations after an earthquake via UAVs. IEEE Access 10:6109–6125. https://doi.org/10.1109/ACCESS.2022.3141216
150. Ho Y-H, Tsai Y-J (2022) Open collaborative platform for multi-drones to support search and rescue operations. Drones 6:132. https://doi.org/10.3390/drones6050132
151. McRae JN, Gay CJ, Nielsen BM, Hunt AP (2019) Using an unmanned aircraft system (drone) to conduct a complex high altitude search and rescue operation: a case study. Wilderness Environ Med 30:287–290. https://doi.org/10.1016/j.wem.2019.03.004
152. McRae JN, Nielsen BM, Gay CJ, Hunt AP, Nigh AD (2021) Utilizing drones to restore and maintain radio communication during search and rescue operations. Wilderness Environ Med 32:41–46. https://doi.org/10.1016/j.wem.2020.11.002
153. Muhamat AA, Zulkifli AF, Ibrahim MA, Sulaiman S, Subramaniam G, Mohamad S, Suzuki Y (2022) Realising the corporate social performance (CSP) of Takaful (Islamic insurance) operators through drone-assisted disaster victim identification (DVI). Sustainability 14:5440. https://doi.org/10.3390/su14095440
154. Redi AANP, Sopha BM, Asih AMS, Liperda RI (2021) Collaborative hybrid aerial and ground vehicle routing for post-disaster assessment. Sustainability 13:12841. https://doi.org/10.3390/su132212841
155. Marturano F, Martellucci L, Chierici A, Malizia A, Giovanni DD, d'Errico F, Gaudio P, Ciparisse J-F (2021) Numerical fluid dynamics simulation for drones' chemical detection. Drones 5:69. https://doi.org/10.3390/drones5030069
156. Fakhrulddin SS, Gharghan SK, Al-Naji A, Chahl J (2019) An advanced first aid system based on an unmanned aerial vehicles and a wireless body area sensor network for elderly persons in outdoor environments. Sensors 19:2955. https://doi.org/10.3390/s19132955
157. García-Cobos FJ, Paniagua-Sánchez JM, Gordillo-Guerrero A, Marabel-Calderón C, Rufo-Pérez M, Jiménez-Barco A (2023) Personal exposimeter coupled to a drone as a system for measuring environmental electromagnetic fields. Environ Res 216:114483. https://doi.org/10.1016/j.envres.2022.114483
158. Restás Á (2022) Drone applications fighting COVID-19 pandemic—towards good practices. Drones 6:15. https://doi.org/10.3390/drones6010015
159. Tubis AA, Ryczyński J, Żurek A (2021) Risk assessment for the use of drones in warehouse operations in the first phase of introducing the service to the market. Sensors 21:6713. https://doi.org/10.3390/s21206713
160. Salazar F, Martínez-García MS, de Castro A, Chávez-Fuentes C, Cazorla M, Ureña-Aguirre J del P, Altamirano S (2023) UAVs for business adoptions in smart city environments: inventory management system. Electronics (Basel) 12:2090. https://doi.org/10.3390/electronics12092090
161. Garcia M, Maza I, Ollero A, Gutierrez D, Aguirre I, Viguria A (2022) Release of sterile mosquitoes with drones in urban and rural environments under the European drone regulation. Appl Sci 12:1250. https://doi.org/10.3390/app12031250
162. Matthews GA (2021) New technology for desert locust control. Agronomy 11:1052. https://doi.org/10.3390/agronomy11061052
163. Wudenka M, Muller MG, Demmel N, Wedler A, Triebel R, Cremers D, Sturzl W (2021) Towards robust monocular visual Odometry for flying robots on planetary missions. In: 2021 IEEE/RSJ international conference on intelligent robots and systems (IROS). IEEE, pp 8737–8744. https://doi.org/10.1109/IROS51168.2021.9636844
164. Granati G, Cichella F, Lucidi P (2021) High-tech training for birds of prey. Animals 11:530. https://doi.org/10.3390/ani11020530
165. McDonnell S, Torcivia C (2020) Preliminary proof of the concept of wild (Feral) horses following light aircraft into a trap. Animals 10:80. https://doi.org/10.3390/ani10010080

166. Al-kabi H, Mazinani SM (2021) DNCS: New UAV navigation with considering the no-fly zone and efficient selection of the charging station. Ain Shams Eng J 12:3669–3676. https://doi.org/10.1016/j.asej.2021.02.020
167. Du L, Li X, Gan Y, Leng K (2022) Optimal model and algorithm of medical materials delivery drone routing problem under major public health emergencies. Sustainability 14:4651. https://doi.org/10.3390/su14084651
168. Abbass MAB, Kang H-S (2023) Drone elevation control based on Python-unity integrated framework for reinforcement learning applications. Drones 7:225. https://doi.org/10.3390/drones7040225
169. Fan Y, Li O, Liu G (2022) An object detection algorithm for rotary-wing UAV based on AWin transformer. IEEE Access 10:13139–13150. https://doi.org/10.1109/ACCESS.2022.3147264
170. Qu Z, Willig A (2022) Sensorless and coordination-free lane switching on a drone road segment—a simulation study. Drones 6:411. https://doi.org/10.3390/drones6120411
171. Rakotonarivo B, Drougard N, Conversy S, Garcia J (2022) Supporting drone mission planning and risk assessment with interactive representations of operational parameters. In: 2022 International conference on unmanned aircraft systems (ICUAS). IEEE, pp 1091–1100. https://doi.org/10.1109/ICUAS54217.2022.9836056
172. Liu C, van Kampen E-J, de Croon GCHE (2023) Adaptive risk-tendency: nano drone navigation in cluttered environments with distributional reinforcement learning. In: 2023 IEEE international conference on robotics and automation (ICRA). IEEE, pp 7198–7204. https://doi.org/10.1109/ICRA48891.2023.10160324
173. Bektash O, Pedersen JN, Ramirez Gomez A, la Cour-Harbo A (2020) Automated emergency landing system for drones: SafeEYE project. In: 2020 International conference on unmanned aircraft systems (ICUAS). IEEE, pp 1056–1064. https://doi.org/10.1109/ICUAS48674.2020.9214073
174. Yuan B, Ma W, Wang F (2022) High speed safe autonomous landing marker tracking of fixed wing drone based on deep learning. IEEE Access 10:80415–80436. https://doi.org/10.1109/ACCESS.2022.3195286
175. Saavedra-Ruiz M, Pinto-Vargas AM, Romero-Cano V (2022) Monocular visual autonomous landing system for quadcopter drones using software in the loop. IEEE Aerosp Electron Syst Mag 37:2–16. https://doi.org/10.1109/MAES.2021.3115208
176. Myeong W, Myung H (2019) Development of a wall-climbing drone capable of vertical soft landing using a tilt-rotor mechanism. IEEE Access 7:4868–4879. https://doi.org/10.1109/ACCESS.2018.2889686
177. Serrano JR, Tiseira AO, García-Cuevas LM, Varela P (2021) Computational study of the propeller position effects in wing-mounted, distributed electric propulsion with boundary layer ingestion in a 25 kg remotely piloted aircraft. Drones 5:56. https://doi.org/10.3390/drones5030056
178. Carlson DF, Akbulut S, Rasmussen JF, Hestbech CS, Andersen MH, Melvad C (2023) Compact and modular autonomous surface vehicle for water research: the naval operating research drone assessing climate change (NORDACC). HardwareX 15:e00453. https://doi.org/10.1016/j.ohx.2023.e00453
179. Bui ST, Luu QK, Nguyen DQ, Le NDM, Loianno G, Ho VA (2023) *Tombo* propeller: bioinspired deformable structure toward collision-accommodated control for drones. IEEE Trans Rob 39:521–538. https://doi.org/10.1109/TRO.2022.3198494
180. Ma M, Xu Y, Wang Z, Fu X, Gui G (2023) Decentralized learning and model averaging based automatic modulation classification in drone communication systems. Drones 7:391. https://doi.org/10.3390/drones7060391
181. Alon O, Rabinovich S, Fyodorov C, Cauchard JR (2021) First step toward gestural recognition in harsh environments. Sensors 21:3997. https://doi.org/10.3390/s21123997
182. Svatý Z, Nouzovský L, Mičunek T, Frydrýn M (2022) Evaluation of the drone-human collision consequences. Heliyon 8:e11677. https://doi.org/10.1016/j.heliyon.2022.e11677
183. Zhu R, Yang Z, Chen J (2022) Conflict risk assessment between non-cooperative drones and manned aircraft in airport terminal areas. Appl Sci 12:10377. https://doi.org/10.3390/app122010377

184. Uddin Z, Altaf M, Bilal M, Nkenyereye L, Bashir AK (2020) Amateur drones detection: a machine learning approach utilizing the acoustic signals in the presence of strong interference. Comput Commun 154:236–245. https://doi.org/10.1016/j.comcom.2020.02.065
185. Morio J, Levasseur B, Bertrand S (2021) Drone ground impact footprints with importance sampling: estimation and sensitivity analysis. Appl Sci 11:3871. https://doi.org/10.3390/app11093871
186. Jeelani I, Gheisari M (2021) Safety challenges of UAV integration in construction: conceptual analysis and future research roadmap. Saf Sci 144:105473. https://doi.org/10.1016/j.ssci.2021.105473
187. Ažaltovič V, Škvareková I, Pecho P, Kandera B (2020) Calculation of the ground casualty risk during aerial work of unmanned aerial vehicles in the urban environment. Transp Res Proc 44:271–275. https://doi.org/10.1016/j.trpro.2020.02.043
188. Lercel DJ, Hupy JP (2023) Exploring the use of geographic information systems to identify spatial patterns of remote UAS pilots and possible national airspace risk. Safety 9:18. https://doi.org/10.3390/safety9010018
189. Lenzi J, Felege CJ, Newman R, McCann B, Ellis-Felege SN (2022) Feral horses and Bison at Theodore Roosevelt National Park (North Dakota, United States) exhibit shifts in behaviors during drone flights. Drones 6:136. https://doi.org/10.3390/drones6060136
190. Sorrell K, Dawlings F, Mackay C, Clarke R (2023) Routine and safe operation of remotely piloted aircraft systems in areas with high densities of flying birds. Drones 7:510. https://doi.org/10.3390/drones7080510
191. Howell LG, Allan BM, Driscoll DA, Ierodiaconou D, Doran TA, Weston MA (2023) Attenuation of responses of waterbirds to repeat drone surveys involving a sequence of altitudes and drone types: a case study. Drones 7:497. https://doi.org/10.3390/drones7080497
192. Frixione MG, Salvadeo C (2021) Drones, gulls and urbanity: interaction between new technologies and human subsidized species in coastal areas. Drones 5:30. https://doi.org/10.3390/drones5020030
193. Thomas K, Granberg TA (2023) Quantifying visual pollution from urban air mobility. Drones 7:396. https://doi.org/10.3390/drones7060396
194. Ahmed M, Cox D, Simpson B, Aloufi A (2022) ECU-IoFT: a dataset for analysing cyberattacks on internet of flying things. Appl Sci 12:1990. https://doi.org/10.3390/app12041990
195. Aldaej A, Ahanger TA, Atiquzzaman M, Ullah I, Yousufudin M (2022) Smart cybersecurity framework for IoT-empowered drones: machine learning perspective. Sensors 22:2630. https://doi.org/10.3390/s22072630
196. Tanveer M, Khan AU, Shah H, Chaudhry SA, Naushad A (2021) PASKE-IoD: privacy-protecting authenticated key establishment for internet of drones. IEEE Access 9:145683–145698. https://doi.org/10.1109/ACCESS.2021.3123142
197. Hamadi R, Ghazzai H, Massoud Y (2023) Reinforcement learning based intrusion detection systems for drones: a brief survey. In: 2023 IEEE international conference on smart mobility (SM). IEEE, pp 104–109. https://doi.org/10.1109/SM57895.2023.10112557
198. Lei Y, Zeng L, Li Y-X, Wang M-X, Qin H (2021) A lightweight authentication protocol for UAV networks based on security and computational resource optimization. IEEE Access 9:53769–53785. https://doi.org/10.1109/ACCESS.2021.3070683
199. Jahan F, Sun W, Niyaz Q (2020) A non-cooperative game based model for the cybersecurity of autonomous systems. In: 2020 IEEE security and privacy workshops (SPW). IEEE, pp 202–207. https://doi.org/10.1109/SPW50608.2020.00049
200. Nalamati M, Kapoor A, Saqib M, Sharma N, Blumenstein M (2019) Drone detection in long-range surveillance videos. In: 2019 16th IEEE international conference on advanced video and signal based surveillance (AVSS). IEEE, pp 1–6. https://doi.org/10.1109/AVSS.2019.8909830
201. Taha B, Shoufan A (2019) Machine learning-based drone detection and classification: state-of-the-art in research. IEEE Access 7:138669–138682. https://doi.org/10.1109/ACCESS.2019.2942944

202. Yang J, Gu H, Hu C, Zhang X, Gui G, Gacanin H (2022) Deep complex-valued convolutional neural network for drone recognition based on RF fingerprinting. Drones 6:374. https://doi.org/10.3390/drones6120374
203. Sun Y, Zhi X, Han H, Jiang S, Shi T, Gong J, Zhang W (2023) Enhancing UAV detection in surveillance camera videos through spatiotemporal information and optical flow. Sensors 23:6037. https://doi.org/10.3390/s23136037
204. Alam SS, Chakma A, Rahman MH, Bin Mofidul R, Alam MM, Utama IBKY, Jang YM (2023) RF-enabled deep-learning-assisted drone detection and identification: an end-to-end approach. Sensors 23:4202. https://doi.org/10.3390/s23094202
205. Ciaburro G, Iannace G (2020) Improving smart cities safety using sound events detection based on deep neural network algorithms. Informatics 7:23. https://doi.org/10.3390/informatics7030023
206. Yang T, Li Z, Zhang F, Xie B, Li J, Liu L (2019) Panoramic UAV surveillance and recycling system based on structure-free camera array. IEEE Access 7:25763–25778. https://doi.org/10.1109/ACCESS.2019.2900167
207. Ajakwe SO, Ihekoronye VU, Kim D-S, Lee J-M (2023) ALIEN: assisted learning invasive encroachment neutralization for secured drone transportation system. Sensors 23:1233. https://doi.org/10.3390/s23031233
208. Alharbi A, Poujade A, Malandrakis K, Petrunin I, Panagiotakopoulos D, Tsourdos A (2020) Rule-based conflict management for unmanned traffic management scenarios. In: 2020 AIAA/IEEE 39th Digital avionics systems conference (DASC). IEEE, pp 1–10. https://doi.org/10.1109/DASC50938.2020.9256690
209. Minucci F, Vinogradov E, Pollin S (2020) Avoiding collisions at any (low) cost: ADS-B like position broadcast for UAVs. IEEE Access 8:121843–121857. https://doi.org/10.1109/ACCESS.2020.3007315
210. Pedro D, Matos-Carvalho JP, Azevedo F, Sacoto-Martins R, Bernardo L, Campos L, Fonseca JM, Mora A (2020) FFAU—framework for fully autonomous UAVs. Remote Sens (Basel) 12:3533. https://doi.org/10.3390/rs12213533
211. Shan L, Li H-B, Miura R, Matsuda T, Matsumura T (2023) A novel collision avoidance strategy with D2D communications for UAV systems. Drones 7:283. https://doi.org/10.3390/drones7050283
212. Kuru K, Pinder JM, Watkinson BJ, Ansell D, Vinning K, Moore L, Gilbert C, Sujit A, Jones D (2023) Toward mid-air collision-free trajectory for autonomous and pilot-controlled unmanned aerial vehicles. IEEE Access 11:100323–100342. https://doi.org/10.1109/ACCESS.2023.3314504
213. Campana I, Bergesio L, Besada JA, de Miguel G (2019) Air tracking and monitoring for unmanned aircraft traffic management. In: 2019 Integrated communications, navigation and surveillance conference (ICNS). IEEE, pp 1–9. https://doi.org/10.1109/ICNSURV.2019.8735284
214. Vagal V, Markantonakis K, Shepherd C (2021) A new approach to complex dynamic geofencing for unmanned aerial vehicles. In: 2021 IEEE/AIAA 40th digital avionics systems conference (DASC). IEEE, pp 1–7. https://doi.org/10.1109/DASC52595.2021.9594499
215. Silalahi S, Ahmad T, Studiawan H (2023) Transformer-based named entity recognition on drone flight logs to support forensic investigation. IEEE Access 11:3257–3274. https://doi.org/10.1109/ACCESS.2023.3234605
216. Chen Y-C, Huang C (2021) Smart data-driven policy on unmanned aircraft systems (UAS): analysis of drone users in U.S. cities. Smart Cities 4:78–92. https://doi.org/10.3390/smartcities4010005
217. Henderson IL (2022) Examining New Zealand unmanned aircraft users' measures for mitigating operational risks. Drones 6:32. https://doi.org/10.3390/drones6020032
218. Koç D, Seçkin AÇ, Satı ZE (2021) Evaluation of participant success in gamified drone training simulator using brain signals and key logs. Brain Sci 11:1024. https://doi.org/10.3390/brainsci11081024

219. Covaciu F, Iordan A-E (2022) Control of a drone in virtual reality using MEMS sensor technology and machine learning. Micromachines (Basel) 13:521. https://doi.org/10.3390/mi13040521
220. Chi NTK, Phong LT, Hanh NT (2023) The drone delivery services: an innovative application in an emerging economy. Asian J Ship Logis 39:39–45. https://doi.org/10.1016/j.ajsl.2023.01.002
221. Jasim NI, Kasim H, Mahmoud MA (2022) Towards the development of smart and sustainable transportation system for foodservice industry: modelling factors influencing customer's intention to adopt drone food delivery (DFD) services. Sustainability 14:2852. https://doi.org/10.3390/su14052852
222. Valencia-Arias A, Rodríguez-Correa PA, Patiño-Vanegas JC, Benjumea-Arias M, De La Cruz-Vargas J, Moreno-López G (2022) Factors associated with the adoption of drones for product delivery in the context of the COVID-19 pandemic in Medellín, Colombia. Drones 6:225. https://doi.org/10.3390/drones6090225
223. Truog S, Maxim L, Matemba C, Blauvelt C, Ngwira H, Makaya A, Moreira S, Lawrence E, Ailstock G, Weitz A, West M, Defawe O (2020) Insights before flights: how community perceptions can make or break medical drone deliveries. Drones 4:51. https://doi.org/10.3390/drones4030051
224. Del-Real C, Díaz-Fernández AM (2021) Lifeguards in the sky: examining the public acceptance of beach-rescue drones. Technol Soc 64:101502. https://doi.org/10.1016/j.techsoc.2020.101502
225. Yaprak Ü, Kılıç F, Okumuş A (2021) Is the Covid-19 pandemic strong enough to change the online order delivery methods? Changes in the relationship between attitude and behavior towards order delivery by drone. Technol Forecast Soc Change 169:120829. https://doi.org/10.1016/j.techfore.2021.120829
226. Tubis A, Werbińska-Wojciechowska S, Sliwinski P, Zimroz R (2022) Fuzzy risk-based maintenance strategy with safety considerations for the mining industry. Sensors 22:441. https://doi.org/10.3390/s22020441
227. Yan R, Dunnett SJ, Jackson LM (2018) Novel methodology for optimising the design, operation and maintenance of a multi-AGV system. Reliab Eng Syst Saf 178:130–139. https://doi.org/10.1016/J.RESS.2018.06.003

Chapter 5
The New Approach to Risk Assessment in Logistics 4.0 on the Example of the Internal Transport Process

Contents

The analysis of the results of the conducted multi-faceted literature review, as well as the results of the own research, clearly indicate that the digital transformation and the implemented technological solutions within Logistics 4.0 require a change in the approach to assessing and monitoring the risks occurring in the cargo handling process within internal logistics and methods to mitigate their occurrence. Research published in scientific articles and conference proceedings in the Scopus and Web of Science databases over the past five years on Warehouse 4.0 and material handling equipment (presented in Chap. 4) indicates that issues relating to the risks associated with the implementation of the cargo handling process within Logistics 4.0 are analysed in narrow, one might even say siloed, areas. In Chap. 2, it was demonstrated that a siloed approach to risk assessment is inefficient and limits the possibility of developing solutions that effectively interact with emerging risks in logistics service systems. Thus, a clear research gap exists regarding an interdisciplinary approach

A. Tubis, *The New Paradigm of Risk in Internal Transport Supporting Logistics 4.0 System*, Lecture Notes in Intelligent Transportation and Infrastructure,
https://doi.org/10.1007/978-3-031-64920-2_5

to assessing risks associated with cargo handling in internal logistics implemented by complex cyber-human-technical systems. Therefore, the research challenge is to develop a risk assessment method that considers the specificities of implementing internal logistics processes handled by cyber-human-technical systems and fills the gap identified in Chap. 4.

5.1 The Internal Transport System as a Cyber-Human-Technical System

Internal transport and warehousing systems form the core of the internal logistics systems that handle material flows within the internal supply chain. Internal transport processes integrate the storage area with the production area and handle material flows along production processes. Their proper coordination makes it possible to balance the demand expressed by production with the warehouse's available materials. At the same time, this coordination makes it possible to eliminate waste associated with excessive stocks or unnecessary movements of transport equipment. Therefore, errors and disruptions occurring in the internal transport process substantially impact the logistics service level and the costs associated with implementing operational processes.

Traditional internal transport systems are human-technical (HT) systems that aim to perform their assigned tasks in a specific space–time. In this system, a human being acts as an operator and decision-maker, while a technical object is an object of human activity designed to perform specific tasks and to which certain characteristics and functions have been deliberately assigned by humans [1]. The HT system (also known as the anthropo-technical system) in engineering sciences is usually represented according to Eq. (5.1) [2]:

$$HTS = \{H, T, M, R\} \tag{5.1}$$

where

$H = \{H_i\}$, $i = 1, 2, \ldots, n$—the set of all operators in the anthropo-technical system;
$T = \{T_j\}, j = 1, 2, \ldots, m$—the set of all technical objects in the anthropo-technical system;
$M = \{M_k\}, k = 1, 2, \ldots, o$—the set of all tasks performed by the anthropo-technical system;
$R = \{R_l\}, l = 1, 2, \ldots, p$—the set of all relations in the anthropo-technical system.

The critical elements in an anthropotechnical system are the human and the technical object, which work together to perform the set of tasks defined for the system. These elements are in a hierarchical relationship with each other, i.e. the superior entity is the operator, who, by controlling the technical object, decides when, where and how to perform the task assigned to them. In the relationship between man and machine, there are manifestations of information processes, but to a minimal extent [2]:

- Through an electronic signalling system, the technical object sends a signal with a specific meaning.
- The human being is the only element in the system that has the ability to receive this signal with the appropriate sense organs. Humans transform this signal into information, based on which the machine operator makes a decision and executes it by acting on the machine's controls, causing it to adjust or stop.

Humans and technical objects also have a relationship with the environment. In this case, the environment is defined as a specific complement of the anthropotechnical system. It is identified as a collection of factors and elements that, on the one hand, influence the system's functioning and, on the other hand, are affected by its functioning. The elements of the environment may be in relationships with the elements of the system of varying importance and strength. The close environment is part of the environment whose influence cannot be ignored and whose impact is often direct. On the other hand, the part of the environment whose influence on the system is indirect and, where justified, can be ignored is referred to as the external environment. It should be emphasised, however, that the physical distance does not determine the qualification as a close or external environment but by the strength of the relationship with the objective function performed by the system [1].

Today's material handling systems are increasingly complex cyber-physical systems that automatically or autonomously handle cargo flows between the warehouse and production systems. Based on the HT system notation presented above (Eq. 5.1), it is possible to describe the CHT system (Eq. 5.2) similarly.

$$CHTS = \{C, H, T, M, R\} \tag{5.2}$$

where

$C = \{C_j\}$, $j = 1, 2, \ldots, q$—the set of all systems providing cyber services and making up the cyber system supporting the operation of the selected process;
$H = \{H_i\}$, $i = 1, 2, \ldots, p$—the set of all persons collaborating within the CHT system (system operators, managers, process participants sharing the workspace);
$T = \{T_k\}$, $k = 1, 2, \ldots, r$—the set of all technical equipment in the CHT system;
$M = \{M_l\}$, $l = 1, 2, \ldots, s$—the set of all tasks (missions) handled by the CHT system in the execution of the selected process;
$R = \{R_m\}$, $m = 1, 2, \ldots, t$—the set of all relations in the CHT system.

Of course, it should be noted that the change introduced is not just adding an extra cyber-system element. In the cyber-human-physical system, the relations between the operator and the technical device, which the cyber-system supports, change. The hitherto hierarchical relationship is replaced by a peer relationship, meaning that the operator ceases to be the master element and begins to work in partnership with the technical device. The mode of communication is also changing. Above all, humans are not the only addressees to the signals sent by technical devices. Thanks to the IoT, data exchange can occur between all elements of the CHT system. Concerning the material handling system, information will be exchanged in a relationship:

- human—transport device,
- human—cyber-system,
- transport device—cyber-system,

but also

- human—human,
- transport device—transport device,
- cyber-system—cyber-system.

The range of possible information connections is shown in Fig. 5.1.

The relationships occurring between the CHT system and the environment are also changing. Primarily, the close environment increasingly influences the correctness of the supported processes. The author's research and the literature review results described in Chap. 4 clearly indicate the critical role of a properly prepared operating environment for automated and autonomous systems such as AGVs, drones and mobile robots. Factors such as the lack of Wi-Fi access, the unevenness of the

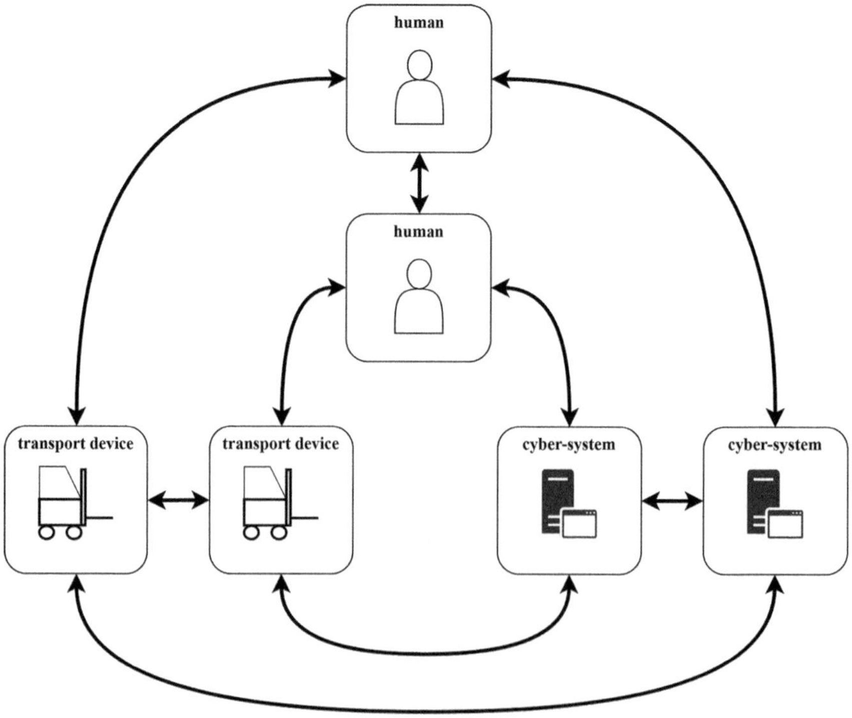

Fig. 5.1 Range of information relationships linking CHT system elements

surface on which the vehicles move, and the lack of proper lighting in the storage area often determine not only the successful completion of the mission but even the complete mission of the autonomous vehicle. Therefore, it is critical to formulate precise requirements for the system's operating conditions, cause-and-effect relationships and the existing relationships between system components under operational conditions.

The external environment also plays a more significant role in the functioning of the CHT system. Never before in traditional human-technical (HT) systems has the development of technological trends been as crucial as it is today. CHT systems not only have to keep up with new technological developments and digital transformation but are generally dependent on actions taken by software providers, including emerging updates. The security of their operation is also constantly vulnerable to cyber-attacks, which can halt the operation of material handling systems and the entire factory. All of this makes it necessary, when assessing the risks associated with the execution of transport processes supported by integrated CHT systems, to consider the impact of the environment, both close and external, on their proper functioning.

However, it should be emphasised that the relationship linking the environment and CHT systems supporting internal transport is two-way, i.e. the systems also interact strongly with the environment. CHT systems are the response of many factories to the need for sustainability. Their positive impact on the environment has been the subject of numerous studies to prove the benefits of implementing CHT systems, such as the company's carbon footprint, reducing waste, or increasing the use of renewable energy. Unfortunately, negative impacts of CHT systems on the environment are also noted, including increased energy demand, technological unemployment and significant changes in the working environment. However, these aspects will not be the subject of an operational risk assessment of the selected process; instead, they may be part of an enterprise-level risk assessment.

5.2 Operational Risk Assessment Method for Cyber-Human-Technical Systems (ORA_CHTS)

The author's many years of research into methods of assessing and managing risks in anthropotechnical systems and the impact of digital transformation and automation on the implementation of logistics processes have led to the development of a new methodological approach to assessing risks occurring in the internal transport process supported by Logistics 4.0 solutions. The basis for the developed method is an interdisciplinary approach to assessing operational risks occurring in logistics processes, which is characterised in Sect. 2.3. This approach combines selected aspects of enterprise risk management and supply chain risk management with an engineering approach focused on the safety and reliability of the operation of technical equipment. Such an interdisciplinary approach to risk assessment in internal logistics is particularly appropriate due to the specificity of internal transport systems constituting a cyber-human-technical arrangement discussed in Sect. 5.1.

The proposed method focuses on assessing operational level risks, as it aims to support the decision-making processes of middle and lower-level managers whose decisions are oriented towards ensuring the continuity of cargo supply and transport service following the adopted logistics service level. At the same time, the method provides analytical tools that continuously support managers' planning processes. The basic assumption for the formulated method is the substantive scope of operational risk, which in this book will be understood as defined in Table 5.1.

According to the above definition, three basic types of adverse events are in the spectrum of interest to analysts, highlighted in Fig. 5.2.

- Type II: Events whose occurrence disrupts the correct course of the process but does not necessarily have a negative impact on the final result achieved. The lack of impact on the achievement of the final results can result from the low impact of the event but also from the activation of contingency plans and safety mechanisms. However, such an event changes how the process is performed, which should be considered in the design of safety systems and contingency planning and, therefore, should be subject to a risk assessment.
- Type III: Events whose occurrence does not interfere with the proper course of the process but limits the achievement of operational objectives—failing to achieve the complete results for the tasks performed. These events follow directly from the definition of risk in ISO 31000:2018, which refers to uncertainty's impact on achieving objectives. These events require special attention in the assessment process, as they can be challenging to detect at the planning or execution stage of the process.

Table 5.1 Definition of operational risk for the interdisciplinary approach (based on [3])

Operational risk means the occurrence of adverse events that disrupt the process's proper course or limit achieving the operational objectives (performance outcomes) adopted for the process.

Fig. 5.2 Matrix of events describing the course of the operational process to the result obtained

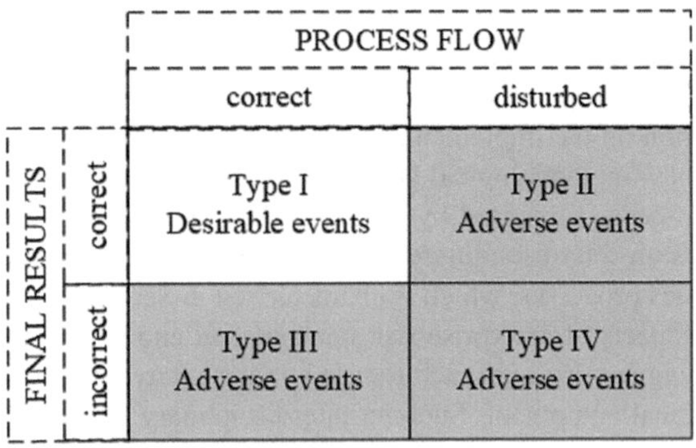

- Type IV: Events whose occurrence disrupts the orderly flow of the process and consequently limits the achievement of operational objectives. This event combines the conditions for occurrence and the effects of the two events described above.

A second important point arising from this definition is that only events whose consequences of occurrence are judged by the decision-maker to be negative, i.e. involving loss or damage, will be the subject of analysis. This approach relates to an engineering view of risk assessment, focused on building a safe working environment to execute efficient and reliable service processes. Thus, managers should not create space to speculate and take "risky games" when managing the material handling process. The required level of safety of the process participants and the reliability of the logistics service provided strongly determines the direction of the analyses, focusing them exclusively on the negative impact of the assessed events.

The proposed risk assessment method is based on good practices identified within the business approach to enterprise risk management. Therefore, a risk management framework has been defined, considering the guidelines of ISO 31000:2018, shown in Fig. 5.3.

In line with the title of this chapter, the results presented in this section of the book focus primarily on the stages of the analytical procedure for the risk assessment aspects, i.e. stages 1 to 4 and the monitoring process (stage 5). In Fig. 5.3, the risk management phase (stage 6) is also included but will not be discussed in detail in this chapter. This risk management stage is still the subject of the author's ongoing research on building resilience in logistics systems and supply chains. Therefore, a detailed characterisation of risk assessment results in the management phase will be described in subsequent publications, continuing the research presented in this book. However, its inclusion in the diagram was necessary to map the overall concept of process risk management and link the assessment results to continuous improvement and process monitoring.

5.2.1 *Identification of the Context*

This step needs to be better recognised in the literature because it is very often treated in a very general way in the studies conducted. Some authors in this step only indicate the area (process) to be covered by the risk assessment and pay little attention to it. In the author's opinion, it is a critical mistake that constitutes a trap for managers who, without an adequately outlined context for the risk assessment to be carried out, receive results that do not match their needs and requirements. Correctly mapping the context for the risk assessment being conducted, as outlined in Fig. 5.4, allows the assessment team to:

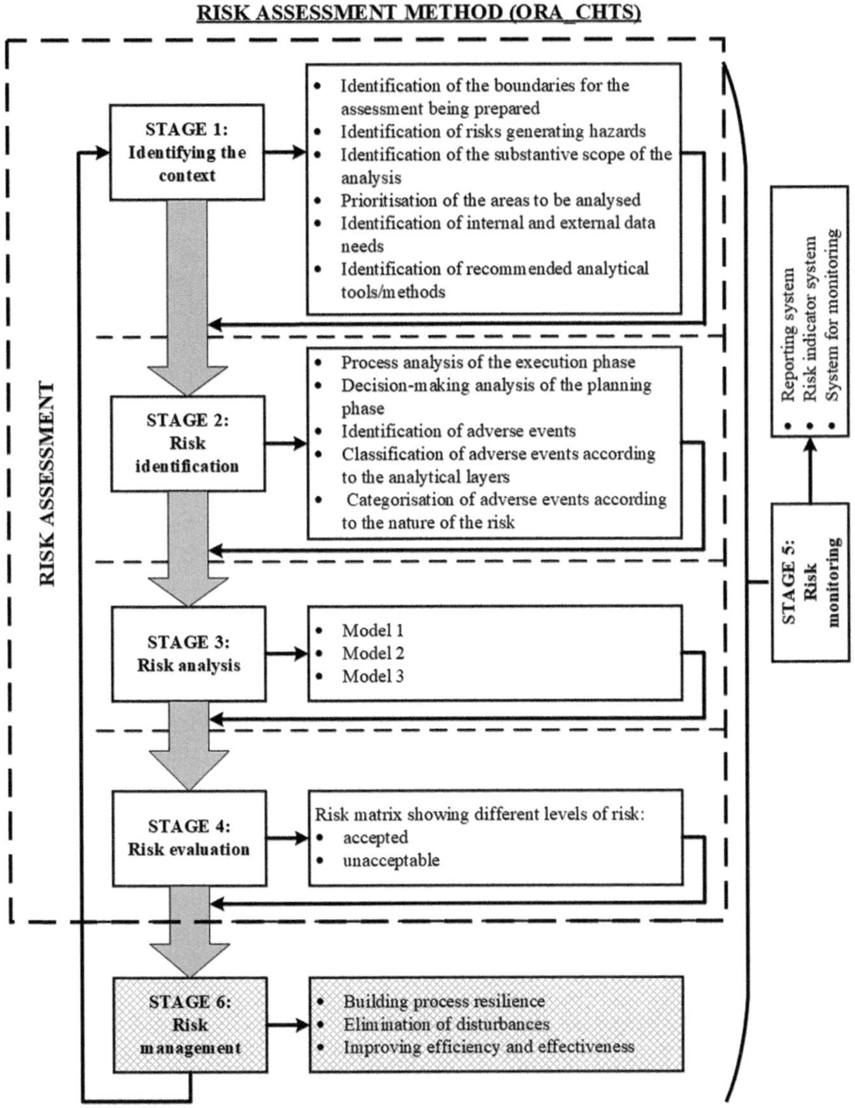

Fig. 5.3 Framework of the operational risk management method for internal logistics processes supported by cyber-human-technical systems

- Unambiguously define the boundaries of the assessment to be carried out, ensuring that the analytical work undertaken is focused on a clearly defined scope of the business.
- Define the substantive scope of the analyses to be prepared well, including the risks generating the risk of adverse events.
- Give appropriate priority to the events being analysed.

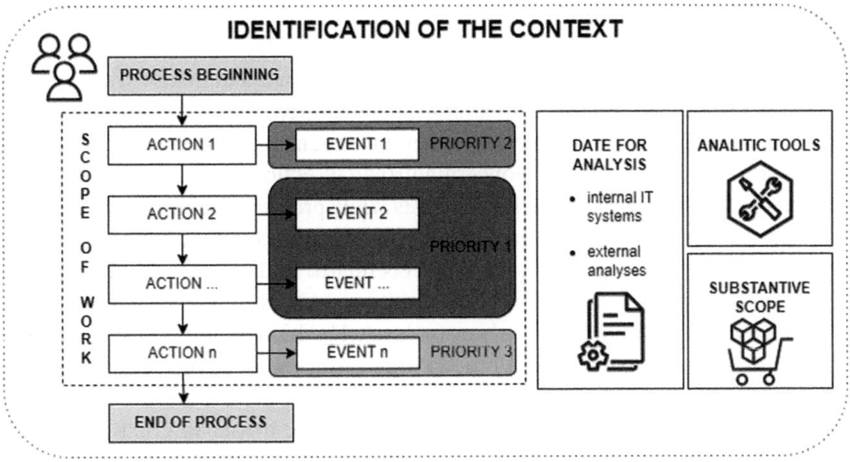

Fig. 5.4 Context of operational risk assessment

- Better identify data needs from internal information systems and external analyses that can be used in risk identification and analysis.
- Identify recommended analytical tools/methods that can be used in the subsequent steps of the risk assessment.
- Therefore, for the proposed operational risk assessment method, an explicit context is proposed for the investigation to be carried out.

Therefore, for the proposed operational risk assessment method, an explicit context is proposed for the investigation to be carried out.

As detailed in Sect. 5.1, the internal transport process in the smart factory is handled by a complex cyber-human-physical system. This system relies on three critical groups of resources actively involved in the execution of the process, posing potential risks to the process and the results achieved. Therefore, outlining the context of risk assessment, it should be pointed out that the subject of the analyses carried out in the Logistics 4.0 system should be adverse events, the occurrence of which results from the activation of a threat, the cause of which may be:

(1) PEOPLE. Unlike traditional transport systems, the source of the activated risk is not only the operator of the transport equipment but also the errors he or she makes. In the case of CHT systems, the scope of the analysis concerning the human resources and their actions, which are the source of the adverse events occurring, is much broader and should include the activity of at least the following work teams: (a) managers planning the cargo handling process within internal logistics; (b) designers of automatic and autonomous systems, responsible for preparing and programming transport equipment for operational work; (c) operators of automatic or autonomous systems, responsible for the daily operation and maintenance of transport equipment and digital systems; (d) other operational staff sharing workspace with transport equipment.

(2) TECHNICAL EQUIPMENT (VEHICLES). As characterised above, the way of working and the role played by technical equipment in the process has changed significantly in CHT systems. These devices move autonomously according to developed algorithms, remaining under limited day-to-day control of the operators. Due to the independence they have been given and even the possibility of self-determination, e.g. concerning route selection, their activity requires a fundamentally new approach to analysis while preparing a risk assessment. Not only do the values of the individual risk measures change, but also the type and structure of emerging adverse events related to their activity. Therefore, to correctly identify a set of adverse events, it is necessary to assess the performance of technical equipment not only as tools used by humans to perform operations but as co-participants in the process that influence its course and the results achieved.
(3) CYBER SYSTEMS. This resource plays a critical role in controlling technical resources and supporting human resource decisions, thus posing a significant hazard to the continuity and correctness of the execution of operational processes. Industry 4.0 has meant that cyber systems have grown from tools supporting the execution of selected operations to becoming a resource controlling the entire process. Realising this new role is particularly important in defining the context for the risk assessment to be carried out, as it will influence the prioritisation of identified adverse events. It is also worth noting that while the effects of a temporary outage of a human or technical resource can be mitigated by introducing a replacement resource into the system (a replacement employee, a replacement vehicle), the effects of even a short-term cyber-system outage cannot be mitigated in this way. This fact makes it necessary to give appropriate priority to incidents caused by digital solutions, depending, of course, on the level of their interference in the implementation of the process (making the course of the process dependent on the efficiency of their operation.

The author's research, the results of which are presented, among others, in [1], also proves the significant influence of the environment on the realisation of logistic processes and the achieved results. The environment in anthropotechnical systems is considered everything outside the human-technical system. The impact of the environment is so significant that, in justified cases, when assessing the risks associated with the operation of anthropotechnical systems, it is recommended to extend the scope of the analysis, which should apply to systems involving anthropo-technical-sphere relations. The internal transport process handled by CHT systems is an example of such a justifiable case, especially as its dependence on the environment is far greater than in traditional systems. Within the defined context of the risk being assessed, it is proposed to distinguish between the close (local) environment and the external environment:

- The analysis of the close environment will refer primarily to the operating conditions and the working space shared by human and technical resources. This implies limiting the impact of the environment to the operational scope related to the flow of handled materials. However, such a limitation is justified from the point of

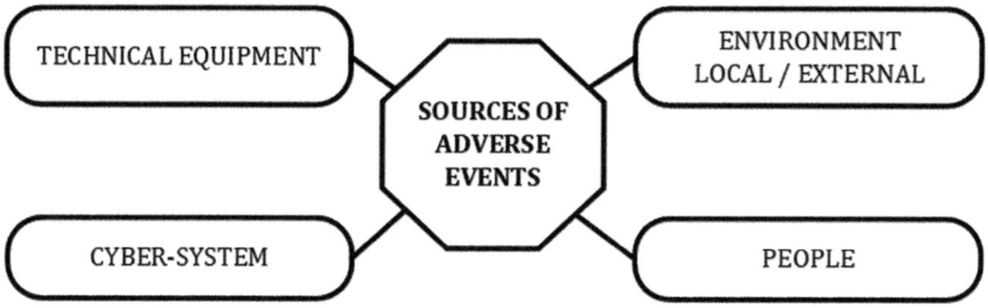

Fig. 5.5 Sources of adverse events occurring in the process of internal transport handled by CHT systems

view of the assumptions and objectives formulated for the proposed assessment method.
- The external environment analysis will focus on assessing the impact of factors and events whose source of occurrence is outside the enterprise.

The range of hazards constituting the source of adverse events affecting the internal transport process is shown in Fig. 5.5.

In order to set the context for the risk assessment, it is necessary to separate the layers of analysis to which the process of identification and analysis of adverse events will relate. Four analytical layers were formulated for the proposed assessment method based on the identified sources of these events. These are:

- human factors layer;
- technical layer;
- digital layer;
- environmental layer.

The second important issue concerning the context of the assessment procedure to be carried out is to define the phases of the process management cycle that the analysis will cover. It will set the boundaries for the risk assessment to be carried out. Three phases of this cycle have been distinguished for formulating assumptions on the proposed approach to operational risk assessment, shown in Fig. 5.6.

Risk analysis can apply to all three phases of the process management cycle, as adverse events can occur in any of them. However, Phase 3 concerns validating the process and the results obtained. Therefore, it is a phase of data collection that forms

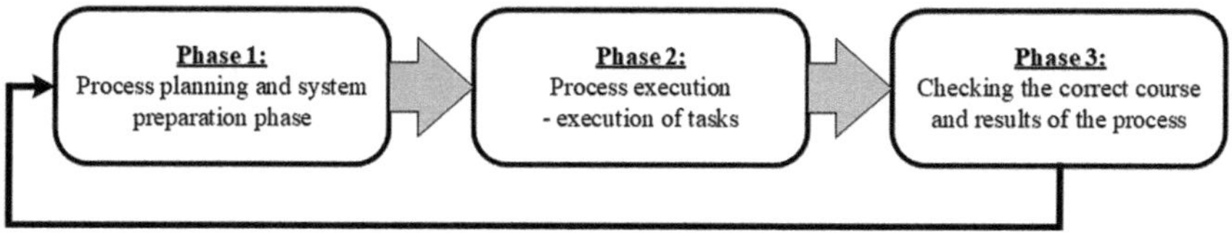

Fig. 5.6 Process management cycle

the basis for the risk assessment to be carried out. Of course, the Author is aware that also in this phase, undesired events may occur, e.g. wrong measurement of the final result, incorrect registration of the deviation from the target value, and lack of registration of information about the occurring disturbances. However, their impact on the correct course of the process and achieving the organisational goals set for the process (according to the adopted risk definition) is indirect. Therefore, in the proposed risk assessment method, this phase will not be included in the analytical procedure, which will only be limited to the first two phases.

The risk analysis should be carried out split into the planning and execution phases of the process. The technical and organisational complexity and the operating rules of CHT systems make the system preparation phase and the operational planning of implementing the assigned tasks of critical importance. This is mainly because the operation process is performed according to a preset algorithm, which is the result of the process of planning and preparing the system for operation. Of course, this algorithm assumes a certain degree of multivariability. However, it is limited to specific, recurring execution scenarios, the need for which is also the result of the planning process. For this reason, it is impossible in CHT systems to follow the old rule of thumb, "plans are there to be changed", which is still used by some companies. It is also impossible to rely on the quick and highly creative reactions of employees who, when an adverse event occurs, take non-standard measures to limit the event's impact. At the same time, the process management cycle phase, to which the conducted risk assessment refers, is decisive at all stages of the analytical procedure—starting with identifying adverse events and ending with the risk management process. At the risk assessment stage, consideration of the process management cycle phase can influence the tools and methods used to identify and analyse adverse events. At the risk management stage, the process management cycle phases determine the choice of potential directions for action and change. If an error occurs in the planning phase, it is possible to influence the probability and consequences of its occurrence. On the other hand, if the disruption occurs in the execution phase, it is only possible to introduce mitigating actions.

Therefore, the recommended context for the risk assessment to be carried out should consider the diagram presented in Fig. 5.7, which reflects the matrix arrangement of the analytical layers considered in the process.

5.2.2 *Risk Identification*

A correctly described context for the assessment procedure to be carried out provides important clues for the next step of the proposed method—identifying adverse events that disrupt the service process or negatively affect the achievement of operational objectives. This step aims to create a structured list of adverse events, considering the source of the risk and its nature. The proposed method recommends adopting a continuous process improvement perspective at this procedure stage. According to this approach, the risk assessment team should focus on unexpected adverse events

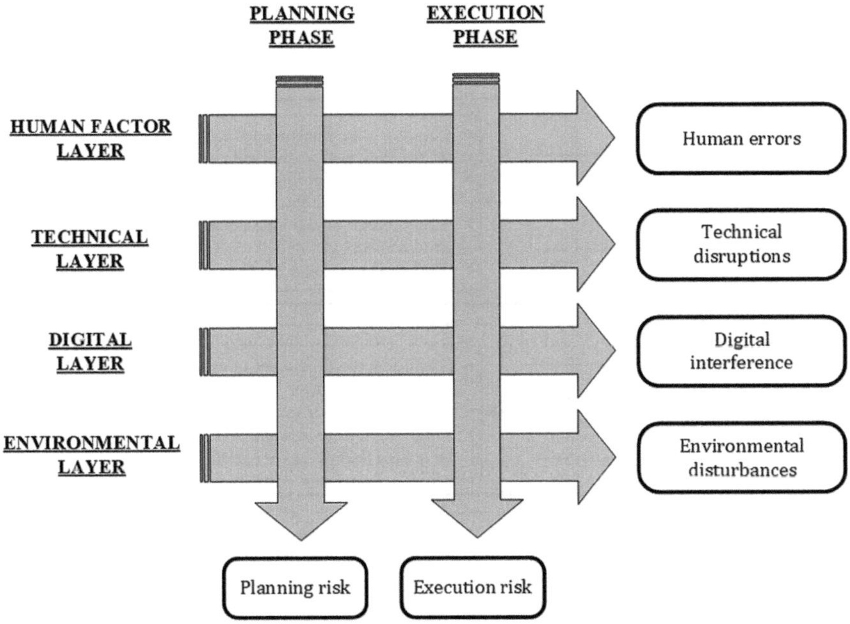

Fig. 5.7 Matrix layout of analytical layers providing context for the proposed risk assessment method

that may suddenly disrupt the process and situations whose effects emerge gradually and negatively impact the process. It should be remembered that certain events do not generate immediate effects when they occur. However, their impact develops in subsequent periods, causing significant losses. The concept of continuous improvement promotes their early identification, even at the initial stage of occurrence. This makes it possible to take pre-emptive risk-reduction measures when the level of risk is still relatively low.

The starting point for identifying adverse events is a procedure that describes the operational process under evaluation, the resources involved, and the objectives adopted due to the actions taken. According to the process-based approach to risk assessment for internal logistics, presented in Sect. 2.3, the criteria for assessing the correctness and efficiency of the operating activities carried out are the time, cost and quality of the process execution, which are characterised in Fig. 2.19. These parameters also determine the outcomes of the process, as the objectives for the internal logistics service are defined. As part of the decision-making processes, managers define target values for each element, considering the trade-off relationships between them.

The risk identification process involves several steps in the analytical process to comprehensively define the set of adverse events to be further assessed. These steps are illustrated in Fig. 5.8.

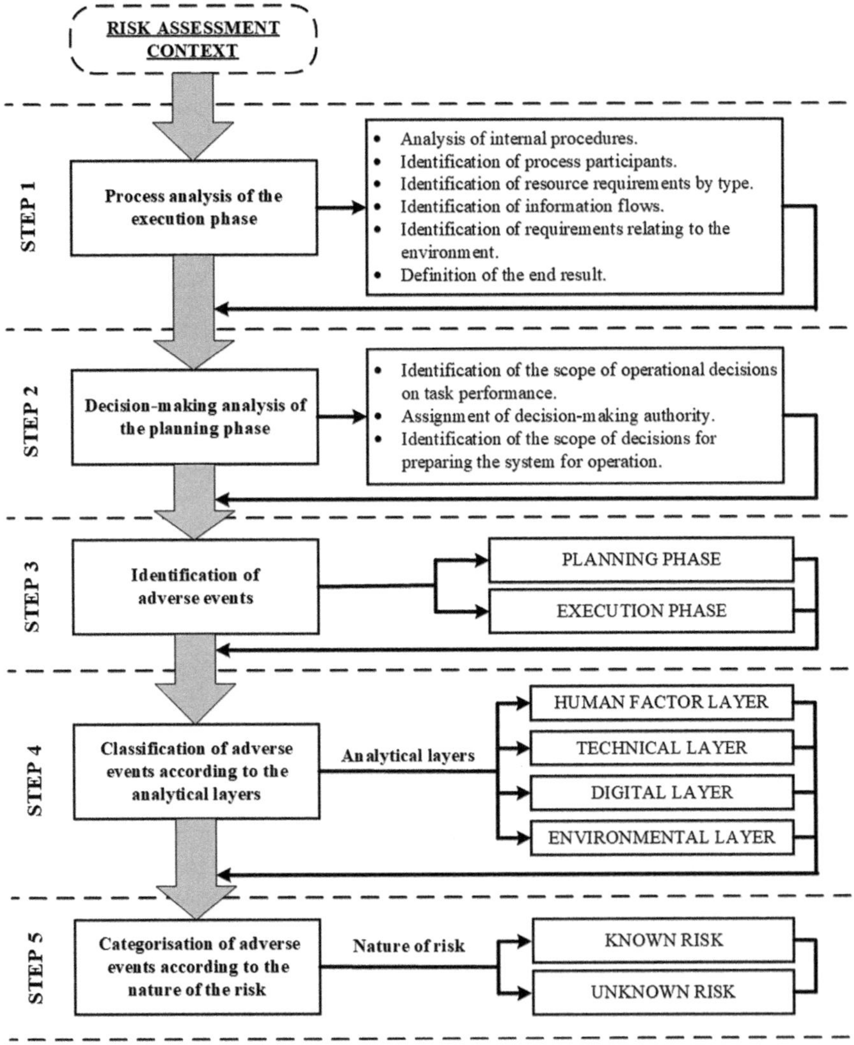

Fig. 5.8 Steps in the risk identification process

The proceeding context defined in the first step implies a two-phase analytical procedure, which should already be mapped in the first assessment step. Therefore, the first two steps concern the decision and process analysis separately. Since the reference point for preparing the operational plans is the model process flow, it is necessary to conduct a process analysis first. This analysis will allow the analysis team to comprehensively reflect the process flow and identify the decision areas that should be the subject of operational planning.

Step 1: Process analysis of the execution phase

The process of identifying adverse events starts with the analysis of the execution phase, as in this step, the handling procedures and the result of the process are defined and form the basis for the planning process. Therefore, swapping the order of the analytical steps in the risk identification process is necessary relative to the order of the process management cycle phases shown in Fig. 5.6.

The process analysis aims to identify the procedures to be followed in the transport service process, the required resources for its implementation and the formulated requirements for the system operating environment. This knowledge is necessary to identify events that may disrupt the standard process flow. When describing an internal transport process (ITP) model, it is necessary to consider all its components relevant to the risk assessment procedure. Therefore, the model will mathematically take the form shown in Eq. (5.3).

$$ITP = \{Pr, ER, As, E, Hz\} \tag{5.3}$$

where

Pr—the procedure adopted. In line with the process approach, this procedure takes into account:

the activities undertaken as part of the internal transport process (A_i), with a clear distinction between the initial activity (A_1) and the final activity (A_m), which identify when the process starts and ends,
the causal and conditional relations ($R_{i\text{-}j}$) that occur, which link these activities into the internal transport service chain.

A model representation of the procedure is given in Eq. (5.4).

$$Pr = \{A_1, A_2, \ldots, A_m\} \cup \left\{R_{1-2}, \ldots, R_{i-j}, \ldots, R_{(m-1)-m}\right\} \tag{5.4}$$

where
$i, j \in \{1, 2, \ldots, m\} \wedge i \neq j$.
ER—the specific final result of the process described by the values defined by the managers for the three process outcomes: cost (C), completion time (T) and completion quality (Q). It is worth noting here that in the process of optimising the final result, a compromise is sought in which it will be possible to achieve a high value of the Q parameter while minimising the values of C and T.

$$\begin{aligned} ER &= f(q, c, t) \rightarrow \max \Leftrightarrow Q = f(q) \rightarrow \max \wedge C = f(c) \rightarrow \min \wedge T \\ &= f(t) \rightarrow \min, \end{aligned} \tag{5.5}$$

As—the resources required to implement the process. According to the adopted evaluation context reflecting the specificity of CHT systems, we distinguish human resources, technical resources (technical equipment) and digital systems among the resources describing the process. However, it is also worth distinguishing the set of information accompanying the implemented activities separately, constituting the information resources describing the process. Since the final result is defined by

$$As = As_1^H, \ldots, As_{k(As)}^H, As_1^T, \ldots, As_{l(As)}^T, As_1^C, \ldots, As_{m(As)}^C, As_1^I, \ldots, As_{n(As)}^I, As_1^F, \ldots, As_{o(As)}^F \tag{5.6}$$

where

As_i^H—human resources, for $i = \{1, 2, \ldots, k(As)\}$.
As_i^T—technical resources, for $i = \{1, 2, \ldots, l(As)\}$.
As_i^C—digital resources, for $i = \{1, 2, \ldots, m(As)\}$.
As_i^I—information resources, for $i = \{1, 2, \ldots, n(As)\}$.
As_i^F—financial resources, for $i = \{1, 2, \ldots, o(As)\}$.

E—the set of requirements for the operating environment of the CHT system, which relate, for example, to the operating conditions under which the transport equipment operates. However, these requirements can refer to the operating conditions of all the distinguished resources: people (e.g. ambient temperature), technical equipment (e.g. visibility of device guidance signs) and systems (e.g. Wi-Fi access).

$$E = \{E_1^H, \ldots, E_{k(E)}^H, E_1^T, \ldots, E_{l(E)}^T, E_1^C, \ldots, E_{m(E)}^C\} \tag{5.7}$$

where

E_i^H—requirements for the working environment of human resources, for $i = \{1, 2, \ldots, k(E)\}$.
E_i^T—requirements for the working environment of technical resources, for $i = \{1, 2, \ldots, l(E)\}$.
E_i^C—digital resource operating environment requirements, for $i = \{1, 2, \ldots, m(E)\}$.

Hz—the set of threats. According to the adopted context of the assessment, the sources of adverse events are human resources (Hz^H), technical resources (Hz^P), digital resources (Hz^C) and the environment (Hz^E). The set of threats is, in this case, the sum of the sets shown in Eq. (5.8)

$$Hz = \left\{Hz^H, Hz^T, Hz^C, Hz^E\right\} \tag{5.8}$$

where

$Hz^H = \left\{Hz_i^H : i = \{1, 2, \ldots, k(Hz)\}\right\}$—the set of human factor hazards.
$Hz^T = \left\{Hz_i^T : i = \{1, 2, \ldots, l(Hz)\}\right\}$—the set of hazards from technical systems.
$Hz^C = \left\{Hz_i^C : i = \{1, 2, \ldots, m(Hz)\}\right\}$—set of hazards from digital systems.
$Hz^E = \left\{Hz_i^E : i = \{1, 2, \ldots, n(Hz)\}\right\}$—the set of hazards arising from the environment, whereby this set consists of the sum of the sets of hazards arising from not meeting the requirements of the working environment (Hz^{LE}) and the hazards from

the external environment (Hz^{EE}). Hence:

$$Hz^{E} = Hz^{LE} \cup Hz^{EE}$$

where the sets: Hz^{LE} and Hz^{EE} are defined as follows:

$$Hz^{LE} = \left\{Hz_i^{LE} : i = \{1, 2, \ldots, n_{Hz(LE)}\}\right\}$$
$$Hz^{EE} = \left\{Hz_i^{EE} : i = \{1, 2, \ldots, n_{Hz(EE)}\}\right\}$$

As defined in this way, the internal transport process model allows us to identify in detail the various types of adverse events that may occur during the execution phase. However, it is worth supplementing this model, presented in mathematical terms, with descriptions detailing specific analytical issues.

The process procedure is a critical element of the analysis, as all other elements that comprise the process model are defined concerning it. It is reasonable to reflect the applicable procedure as a process map for process analysis. In order to prepare such a map, various mapping tools are used, which allow it to reflect the activities performed within the procedure along with the relations between them. Simple process diagrams allow us to map the steps described in the procedures and determine the order in which they are carried out. In this way, they meet basic analytical requirements. However, for risk assessment purposes, they need to be complemented to assign appropriate roles to the process participants and enrich the diagram with information flows. An example of such a complementary tool is the RASCI matrix (Responsible, Accountable, Support, Consult, Inform) [4]. It is a matrix that allows assigning appropriate roles to individual process participants in the form of a matrix. The tasks performed within the process are entered in the matrix rows, while all process participants are identified in the columns. At the intersection of the rows and columns, an appropriate symbol is entered to denote the role played by a given participant in performing a given task. The RASCI matrix is a particularly useful tool for mapping the roles played by CHT system components in handling the material handling process. The unambiguous assignment of responsibility for the performance of individual tasks reduces the possibility of conflicts and disruptions in human-technical object cooperation. From a risk assessment point of view, the analysis of the RASCI matrix will identify the points of contact between human-technical object, human-digital system, and technical object-digital system cooperation, which may cause adverse events. A more advanced process mapping tool can be BPMN (Business Process Model and Notation), which reflects in its notation both the entities involved in the process and the sequence of actions taken with their associated information flows. BPMN notation is recommended for mapping processes, which are written as algorithms controlling cyber-physical systems.

A comprehensive process analysis makes it possible to determine the demand for resources necessary to perform individual tasks in terms of type (indication of the type of resource used in the performance of a given activity), as well as the demand for information accompanying material flows and necessary for the performance of individual tasks. Determining the need for resources used in activities is the basis for assessing their availability and regularity, which may be the cause of occurring adverse events. At the same time, the need to identify the sets of information accompanying the performance of each activity should be emphasised. Distinguishing the sets of information collected and exchanged in the execution of the internal transport process as a separate resource is justified from the point of view of the risk analysis carried out. In addition to the systems for collecting, analysing and distributing data, the distinguished digital resources may include the data itself, but only in digital form. Meanwhile, as part of the execution of the process, certain information may still be collected or exchanged traditionally. Of course, it only applies to messages exchanged between people. Identifying such data sets and communication links resulting from the procedure in place is particularly important in identifying potential adverse events.

The final result defined for the internal transport process is the basis for the planning process. On this basis, resource requirements are quantified. Unambiguously defined result parameters also provide a reference point for assessing the impact of occurring incidents.

Step 2: Decision-making analysis of the planning phase

The adopted context for the analyses carried out assumes that the process of identifying adverse events should differentiate the disruptions, the occurrence of which is characteristic of each of the process operation phases. At the same time, however, the process analysis and the final result parameters identified are the basis for the operational planning process. The results of the analysis made it possible to determine the scope of operational decisions regarding implementing assigned tasks. Their identification, together with the assignment of decision-making powers to managers and cyber-systems, is critical from the point of view of identifying adverse events. Mistakes made by managers at the operational planning stage can generate critical disruptions in the process, as well as negatively affect the achievement of the intended final results of the process. The mistakes may be due to a lack of access to the required information or a lack of competence to make certain decisions. For this reason, it is necessary to identify the following as part of the decision-making analysis:

- the competence (knowledge and experience) of the decision-makers;
- the need for information to support the decision-making processes of the measures;
- the sources of this information;
- the need for quantitative and qualitative analyses to support decision-making processes;
- the costs of obtaining this information.

At the same time, some decisions may be taken automatically by the cyber-system. In this case, faulty operational planning may result from lacking the required up-to-date data feeding the digital analytical systems and faulty analytical algorithms implemented in digital solutions. Determining the decision scopes assigned to the cyber-system will allow better identification of the causes of the distortions occurring in the process.

An essential element of the decision analysis is the definition of the required scope of system readiness. This scope defines the activities that precede the process execution phase and concerns ensuring that the necessary resources identified in the process analysis are available and ready to work. At this stage, the resources are planned quantitatively, which is necessary to achieve the defined final results. This analysis stage also concerns decisions related to creating a suitable working environment. It is therefore necessary to determine which decisions regarding the operating environment and shared workspace should be made in the planning phase and how system readiness is confirmed.

Step 3: Identification of adverse events

The traditional approach to risk assessment in human-technical systems focuses primarily on the execution phase. When identifying adverse events, analysis teams primarily look for the answer to the question "What could disrupt the system?". It should be noted that this question naturally relates to the system's operation during the performance of assigned tasks. Of course, a cause-and-effect analysis sometimes reveals the impact of erroneous decisions made during the planning phase. The results of these analyses then make it possible in the risk management phase to improve decision-making processes in the company. However, it is not common practice to consider already at the adverse event identification stage what planning errors may affect process execution. However, such an approach seems necessary in the case of CHT systems. Above all, the consequences of errors made in the task planning and system preparation phases are disproportionately higher than in traditional transport service systems. In human-technical systems, the operator is the decision-maker and can make decisions at any time regarding the activity carried out by himself and his subordinate machine. In the event of a disruption caused by an error in the planning phase, the operator often took non-standard measures to mitigate the effects of erroneous decisions. In the case of CHT systems, a planning error often means that the entire system has to be stopped, and a required correction has to be made to the prepared algorithm for the planned activity. These actions have the effect of increasing the execution time of the process and additional operating costs. The process of preparing the system for operation is also critical. As demonstrated in Sect. 5.1, a high level of integration of technical, digital and human system operations requires that the system components are coordinated and work together. The lack of readiness of one of these elements can bring the whole system to a temporary halt. While replacement of staff or transport equipment is possible, the lack of operability of the digital system is already a critical situation. For this reason, it makes sense for the risk assessment procedure to focus on possible errors made in planning tasks to prepare the system for operation. Therefore, the proposed change regarding the two-phase approach to identifying adverse events is important and necessary in terms of the quality of the results obtained from the developed risk assessment method.

The results of the process analysis of the execution phase and the decision analysis of the planning phase allow the creation of a knowledge base about the internal transport process. This knowledge forms the basis for identifying adverse events, which should reflect the disturbances occurring in both phases of the transport service. Since adverse events, by their characteristics, are strictly assigned to the process management cycle phase, analytical techniques appropriate to the scope of activities undertaken in each of the two phases should be used to identify them. This means that when assessing the human factor layer in the planning phase, incorrect human resource planning methods should be identified, and in the implementation phase, human errors committed should be identified. A similar approach should be adopted for the technical layer. In the planning phase, errors relating to the planned intensity of use and planned maintenance operations will be identified. In contrast, in the implementation phase, risk identification will refer to the occurrence of damage.

A list of adverse events is created for further risk analysis based on the identification procedure performed.

Step 4: Classification of adverse events according to the analytical layers

The next step is to group the adverse events according to the defined analytical layers in Fig. 5.7. This classification will allow analysts to understand better the root causes of the adverse events that occur, translating into more efficient and effective risk management in the mitigation and monitoring phases. By assigning a root cause to each event, it will be possible to select effective and efficient ways of dealing with the risk, which can precisely address the source of occurrence.

The classification also aims to identify events whose occurrence is generated by more than one cause. Their identification in the list of adverse events is critical regarding the effectiveness of measures taken to reduce the likelihood of their occurrence. Lack of awareness of the complexity of the causes of occurring risks may result in high expenditure on risk reduction, the effectiveness of which will be lower than expected. At the same time, multiple causality may indicate that the source of an adverse event is the relationship between the selected resources rather than the resources themselves. Such knowledge is also crucial for further risk management, as correctly identifying cause-and-effect relationships is an important element of success in implemented solutions.

Step 5: Breakdown of adverse events according to the nature of the risk

The final step in this assessment is categorising the identified adverse events into known and unknown risks. This classification is critical for selecting analysis methods in step 3 of the proposed operational risk assessment method. As characterised in Chap. 2, the following assumptions are made for the classification performed:

Known risks will comprise a set of adverse events that can be identified and measured based on quantitative and qualitative analyses. The probability and consequences of these events can be calculated from historical data or estimated based on expert knowledge. For this group, tools and methods aimed at reducing the probability and mitigating the impact of the risk can be applied at the risk post-treatment stage.

	PLANNING PHASE		EXECUTION PHASE	
HUMAN FACTOR LAYER	AE – H/PP/K	AE – H/PP/UK	AE – H/EP/K	AE – H/EP/UK
TECHNICAL LAYER	AE – T/PP/K	AE – T/PP/UK	AE – T/EP/K	AE – T/EP/UK
DIGITAL LAYER	AE – D/PP/K	AE – D/PP/UK	AE – D/EP/K	AE – D/EP/UK
ENVIRONMENTAL LAYER	AE – E/PP/K	AE – E/PP/UK	AE – E/EP/K	AE – E/EP/UK
	KNOWN	UNKNOWN	KNOWN	UNKNOWN

Fig. 5.9 Classes of adverse events that form the basis of the risk classification exercise

Known risks will comprise a set of adverse events that can be identified and measured based on quantitative and qualitative analyses. The probability and consequences of these events can be calculated from historical data or estimated based on expert knowledge. For this group, tools and methods aimed at reducing the probability and mitigating the impact of the risk can be applied at the risk post-treatment stage.

The result of the identification exercise should be a list of adverse events for which a risk analysis will be carried out. However, the events on this list should be grouped according to the accepted risk assessment context and the proposed classification criteria formulated in the adverse event identification exercise. In this way, 16 possible classes of adverse events are distinguished, as shown in Fig. 5.9

5.2.3 *Risk Analysis*

The proposed assessment method adopts the classical way of expressing risk as the product of a measure of the frequency of an adverse event and a measure of the consequences of its occurrence. Thus, the risk ratio (RPN) will be estimated using the formula (5.9).

$$RPN = O \times C \tag{5.9}$$

where

O—a measure of the frequency of occurrence of an adverse event,
C—a measure of the consequences of the adverse event.

How the frequency of an adverse event is determined depends primarily on the nature of the risk being assessed and the data on the CHT system collected in the company's knowledge bases. The extent and quality of the data available to the risk assessment team depends mainly on the reporting system, which should be dedicated to the assessed process and the supporting CHT system. The broader the reporting of data on adverse events and their accompanying conditions, the greater the accuracy of the risk analysis performed and the precision of the results obtained. If an organisation collects data continuously and comprehensively according to the information needs of the risk assessment team, then estimating the frequency of occurrence of an adverse event will be possible based on a probability distribution describing the random occurrence of the event. However, if the company does not collect all adverse event data, then the determination of risk parameters can only be based on the experience and knowledge of management and the risk assessment team.

In the proposed risk assessment method, estimating the consequences of an adverse event according to the concept of vulnerability of the process to the identified risks is recommended. The concept of process vulnerability is presented in [3]. Vulnerability is assessed by looking at the impact of an event on defined process evaluation parameters. It is worth noting that the assessment concerns the impact's effect (e.g., a reduction in the service's quality) and the impact's duration (how long the indicator is below the target value). The impact assessment of an adverse event according to the proposed approach is shown in Fig. 5.10.

According to Fig. 5.10, the proposed approach should assess:

- the direct and indirect effects of the adverse event on the achieved process outcomes (e.g. increased handling time of a transport order, reduced level of customer logistics service);
- the duration of the impact of the identified adverse events on the internal transport process in subsequent periods.

The assessment of the impact of adverse events from a short-term perspective of the transport order handling process refers to three parameters describing the final result of the process. The assessment of the long-term consequences of the occurrence of the event, on the other hand, refers to the cost-quality parameters recorded from the organisational level. Estimating the consequences of an event thus refers to the consequences borne by the company at the time of the event and in future periods. Therefore, these consequences can take the form of:

- a non-recurring cost that is directly related to the occurrence of the adverse event (attributable to the process); these can be actual costs (e.g. additional pay for employee overtime), including the costs of extraordinary events (e.g. the cost of repairing a damaged AGV);
- costs arising in the future, which are not directly recorded in the accounts but which may arise and even accrue in subsequent periods as a result of the event (e.g. loss of credibility that will reduce demand for the company's products and

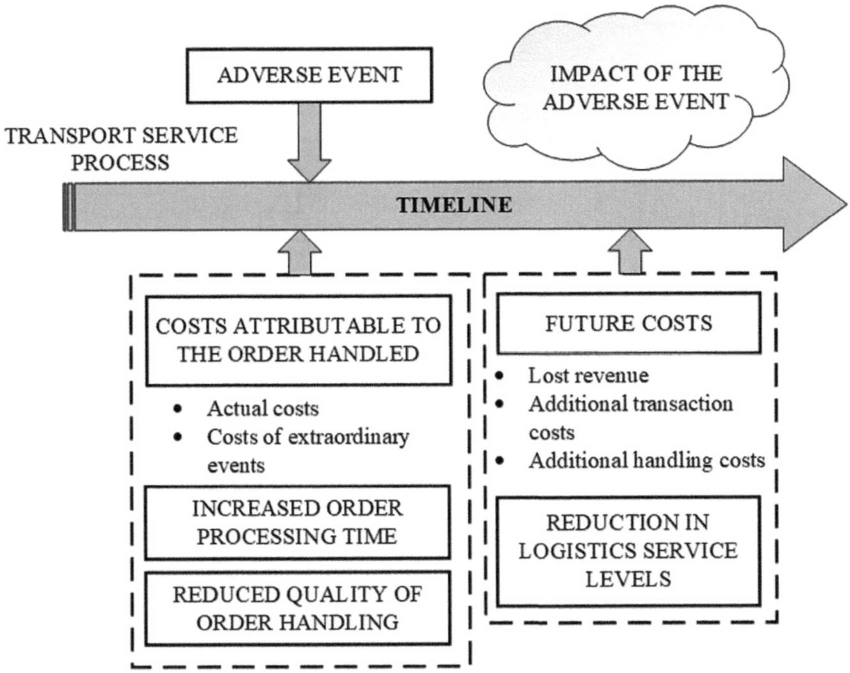

Fig. 5.10 Assessment of the impact of adverse events in the short and long-term (based on [3])

loss of potential revenue, or higher maintenance costs for transport equipment). These costs result from the extended activation time of the adverse event's effects.

The prepared adverse event matrix shown in Fig. 5.9 is the basis for the risk assessment. The resulting division into known and unknown risks is essential at the analysis stage. This division aligns with the definition presented in step 5 of the risk identification and determines the analytical team's level of knowledge about the investigated events. On this basis, it is possible to select effective risk estimation techniques (appropriate to the information resources available). Figure 5.11 presents recommendations for selecting risk analysis models depending on the organisation's knowledge of the phenomenon under investigation.

Model 1: Risk analysis based on historical data on previously recorded adverse events

The higher the level of knowledge about the adverse events analysed in an organisation, the more accurate the risk index estimation associated with their occurrence. The greater the collection of historical data on the frequency and impact of adverse events, the more precise the analytical methods a company can use. A highly valued and expertly recommended way of assessing risk is to use quantitative analytical methods. Determining the value of a risk indicator based on the use of quantitative methods increases the reliability of the presented results for decisions at the

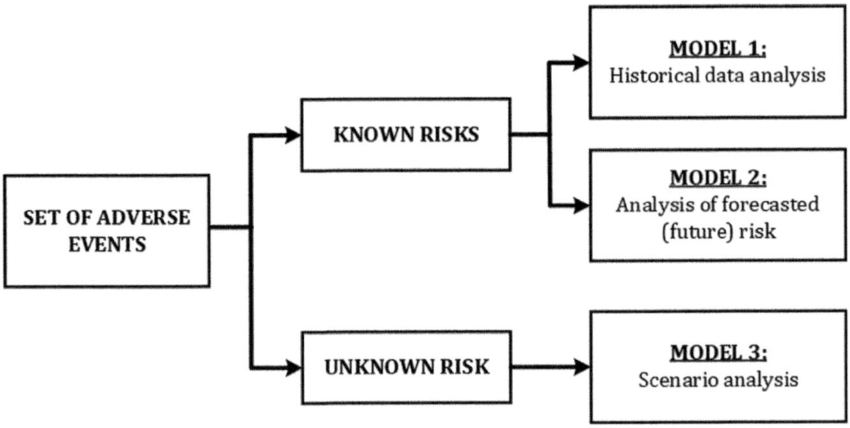

Fig. 5.11 Selection of risk analysis models

operational level. Quantitative methods dedicated to operational-level risk analysis are usually divided into statistical and operational research methods. Examples of methods from both groups are shown in Fig. 5.12.

However, it should be remembered that the quality of the results is influenced not only by the method of analysis used but also by the quality and availability

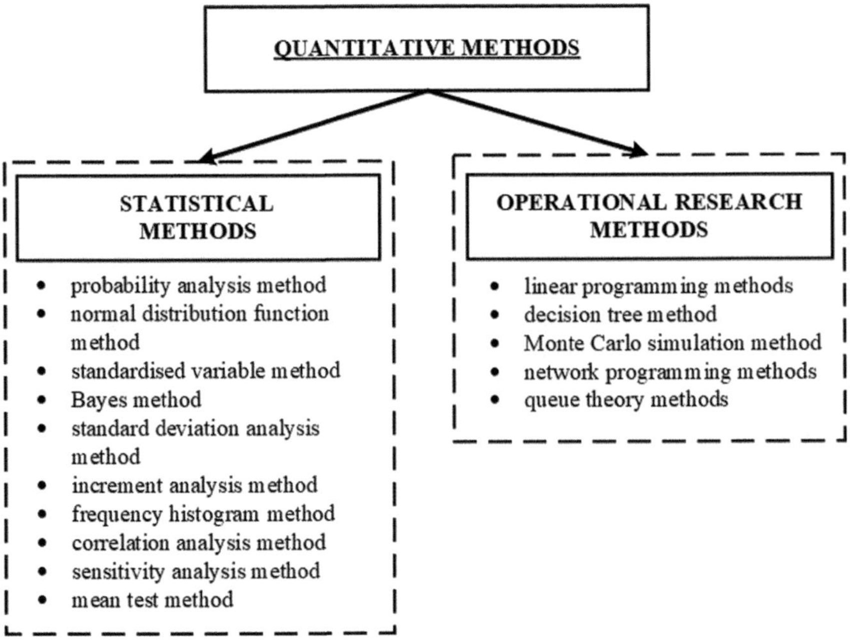

Fig. 5.12 Quantitative methods dedicated to operational-level risk analysis (based on [3])

of the required data. Collecting and updating data on disruptions occurring and associated losses is a critical challenge for this group of risk assessment methods. Quantitative methods require access to up-to-date data on events in the company and the environment. Therefore, it is necessary to define a set of internal and external data that can effectively feed analytical systems. The more significant organisational challenge is the acquisition and collection of external data. Firstly, it is necessary to clearly specify which data (phenomena) should be recorded in the organisation's knowledge bases.

Primarily, this should be data relevant to understanding the risks to which the company is exposed. Therefore, the starting point for acquiring these data should be the results of the conducted proceedings identifying adverse events, especially those belonging to the AE-E group, i.e. those in which the source of risk is the environment. Identifying these events and collecting data to estimate the risk of their occurrence is critical since, in most cases, the company has limited or no influence on activating this risk. The second critical challenge regarding external data is its availability. Since they relate to phenomena that can be monitored by the enterprise with a specific time lag or cannot be monitored at all because they relate to the activities of other entities, it is almost impossible to access detailed knowledge about them.

The collection of internal data seems to be a more straightforward task. Many researchers emphasise that information systems currently being developed as part of the digital transformation can strongly support this process. In CHT systems, the support of data collection by cyber systems even seems natural. Indeed, it is, but only on the condition that a data reporting system is set up that is adapted to the requirements of the risk analyses being prepared. Therefore, the data must be reported in an appropriate structure and updated at the required frequency. Research conducted by the Author has confirmed, for example, that in risk analysis, it is more beneficial to aggregate data according to the modal value (dominant) of a particular phenomenon than according to the average value [3]. Often, the information needs of the risk assessment process also necessitate an extension of the scope of the reported data. Indeed, for analysis, it is necessary to collect data not only on the consequences of a given adverse event but also on the conditions accompanying it. This is particularly important in CHT systems, where an adverse event is often associated with a combination of several factors leading to the activation of the hazard. Recording these factors is critical, mainly when digital systems use the data collected for the "learning" process. Limited data recorded in the system are responsible for an incomplete or incorrect view of the phenomenon in question, resulting in erroneous conclusions from the analyses performed.

For CHT systems, there is another challenge related to using quantitative methods to estimate known risks based on collected historical data. The result of the development of Industry 4.0 and digital transformation is the increasing number of implementations relating to cyber-physical systems supporting internal processes in organisations. However, both phenomena are still at an early stage of development in many companies. This means these organisations do not yet have enough data or shaped experience curves to ensure that the required knowledge assets are available at the right level. For this reason, quantitative–qualitative analysis methods are often

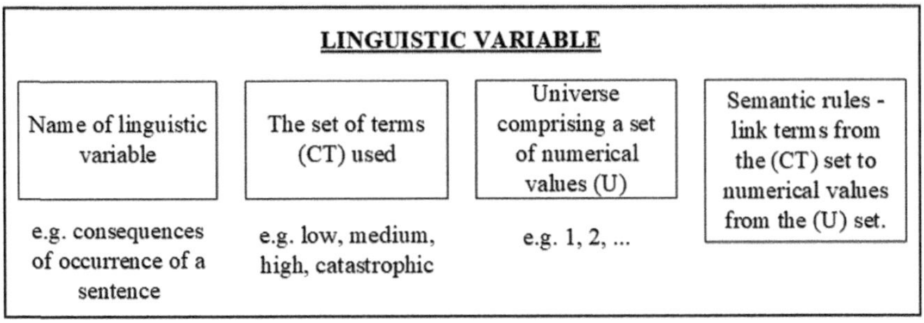

Fig. 5.13 Description of the linguistic variable (based on [6])

used for risk analysis, which allows risks to be estimated even without the required availability of historical data, e.g. based on expert opinion.

In the case of limited access to quantitative data, a frequently used approach is a linguistic approach to estimate the magnitude of parameters describing risk. This approach involves using verbal (linguistic) variables and the relationships between them [5]. The primary concept in this case is the linguistic variable (linguistic variable), whose value is not a number but simple words or sentences used in natural language in human communication. The four elements shown in Fig. 5.13 describe the linguistic variable.

Analyses based on a linguistic approach use the knowledge and experience of experts or employees (mainly managers) in situations where the company does not have the historical data collected to benefit from a quantitative approach.

Linguistic variables are used, among other things, in fuzzy set analysis. This concept was first applied by Zadeh [7] for situations in which the decision-maker is faced with uncertainty and imprecision in the information used in the analytical process. The theory is a valuable analytical tool when measuring ambiguous concepts linked to people's subjective judgements [8]. Linguistic variables are used in this case to express experts' opinions regarding evaluating the two parameters describing the estimated operational risk. According to fuzzy set theory, the risk parameters for each adverse event are treated as a fuzzy number (FN) to which a specific membership function (MF) is assigned. It is worth noting here that the membership functions can take different shapes, with the choice of shape being a subjective decision and depending on the issue being analysed. In both theory and practice, triangular or trapezoidal functions are most commonly used [8]. The triangular fuzzy number is represented by the triangle $A_Z = (a, b, c)$, while the membership function will be described by the formula (5.10).

$$\mu_z(x) = \begin{cases} 0 \quad for\ x < a \\ \frac{x-a}{b-a}\ for\ a \le x \le b \\ \frac{c-x}{c-b}\ for\ b \le x \le c \\ 0 \quad for\ x > c \end{cases} \tag{5.10}$$

In the membership function described in this way, the parameters a and c denote the lower and upper bounds for the described fuzzy number A_Z, while the parameter b indicates the modal value of A_Z.

If a trapezoidal fuzzy number is used, it will be described as $A_Z = (a, b, c, d)$, and its membership function will be expressed by the formula (5.11).

$$\mu_z(x) = \begin{cases} 0 & for\ x < a \\ \frac{x-a}{b-a} & for\ a \le x \le b \\ 1 & for\ b \le x \le c \\ \frac{d-x}{d-c} & for\ c \le x \le d \\ 0 & for\ x > d \end{cases} \quad (5.11)$$

In the membership function described this way, the parameters a and d represent lower and upper bounds for the described fuzzy number A_Z.

Although triangular and trapezoidal functions are most commonly used in fuzzy set theory for risk assessment, the Author recommends performing tests to confirm the adequacy of the chosen function before making a decision. It is worth noting that Gaussian FNs were best suited to the issue under study in one of the studies conducted by the Author [7] concerning the risk assessment of using drones in warehouses based on fuzzy sets.

Based on the research, the Author also recommends using the MAMDANI model, which consists of four modules: Fuzzification, Knowledge base, Fuzzy Interference System and Defuzzification [9]. The analytical process following the MAMDANI model is shown in Fig. 5.14.

The analytical procedure involves mapping fuzzy inputs and controlling outputs according to fuzzy set theory. Accordingly, the knowledge base consists of [9]:

- a database in which the membership functions of the fuzzy sets used in the rules applied are defined and
- rule bases containing 'if–then' rules, which are used to capture imprecise ways of reasoning.

The Fuzzy Interference System uses the MIN and MAX operators. The MIN operator is used for operations involving combinations and implications, while MAX is used to aggregate fuzzy results. The final stage of the procedure is defuzzification, which aims to convert the fuzzy output into a crisp output signal. The output value is determined by Eq. (5.12). [10]

$$\text{Centroid of area, } z^* = \frac{\int \mu_A(z) \cdot zdz}{\int \mu_A(z)dz} \quad (5.12)$$

where z^* is the defuzzified output, and $\mu_A(z)$—the aggregated output member function.

After defuzzification, the fuzzy inference system returns a crisp output value that explicitly expresses the risk level of the adverse event under analysis.

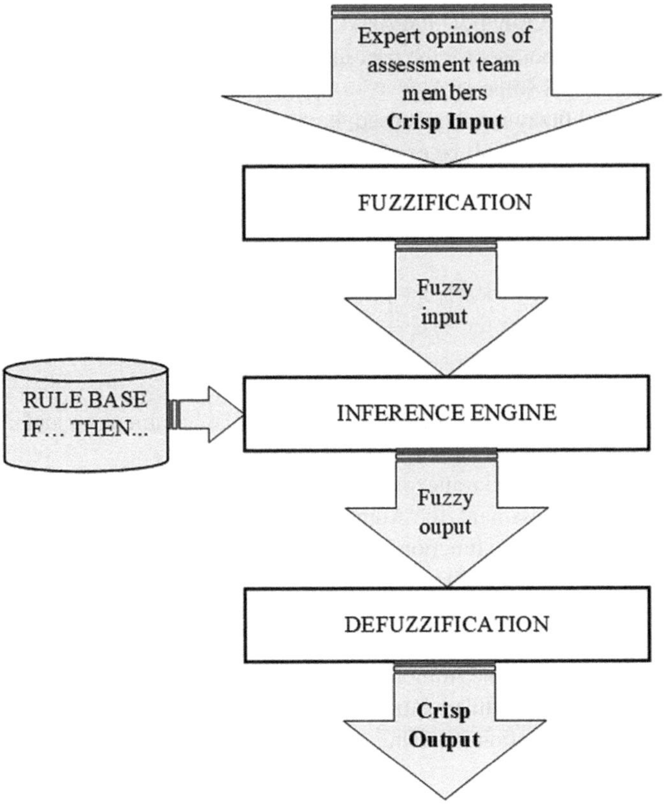

Fig. 5.14 The MAMDANI model in fuzzy set analysis (based on [9])

Model 2: Risk analysis of future (predicted) adverse events—known but not yet recorded in the system

Risk analysis based on historical data is limited to events that have already been recorded in the enterprise and about which the organisation's knowledge can be expressed in the form of quantitative data. However, as highlighted above, CHT systems are new solutions only being implemented in enterprises and regularly developed as the enterprise gains further levels of digital maturity. Therefore, it is difficult to assess the risks associated with their operation comprehensively, as many adverse events may not yet have occurred during their initial periods of operation. An example of such an event is battery ageing, which manifests itself in an increasingly short phase of use between maintenance. In the initial phase of operation, such a risk hardly occurs with properly functioning batteries and equipment, whereas it increases with the lifetime of the transport equipment.

For this reason, CHT systems also require a second risk analysis model, which refers to identified adverse events with no history of occurrence in the company under study. These events are known to managers because they have been recorded in other companies, among supply chain partners or in similar CHT systems. Experts also have knowledge of these events and can actively participate in the analysis process. In analysing these events, data from the environment on adverse events recorded by others play a unique role. The problem in this case is the availability of this data, which is usually the internal data of other organisations unwilling to share it.

The solution to this problem is for companies to work together in Supply Chain Risk Management (this concept is characterised in Chap. 2). Within this concept, it is possible to share knowledge about the frequency and impact of adverse events in the various phases of the life cycle of cyber-physical systems with partners with more experience. In addition, it is possible to analyse the strength of the impact of an occurring event on the activities of other participants in the supply chain, which makes it possible to assess its effects not only on a micro scale (company) but also on the scale of all cooperating links (estimation of the effects of the occurring domino effect.

A second possibility for acquiring the required knowledge is cooperating with research units that conduct analyses in this area. This cooperation can take the form of:

- active, e.g. through joint research projects that evaluate potential adverse events based on observations, simulations or experiments;
- semi-active, e.g. through expert panels in which representatives of research institutions act as experts to assess the likelihood and consequences of identified adverse events;
- passive, e.g., through sharing research reports and scientific publications that describe the results of analyses for similar CHT systems, which can increase the organisation's knowledge of the phenomena under investigation.

A third possibility is the use of expert panels, which can include representatives not only from science but also from industry. The opinions gathered in a group of experts can form the basis for estimating risk indicators based on fuzzy set theory, which was characterised when describing Model 1. Due to the lack of historical data, using this method is practically the only way to map the level of risk using numerical variables.

If it is not possible for a company to use the above-described means of obtaining knowledge of forecast adverse events, it should use qualitative methods to assess future risk. These use subjective assessment measures but are characterised by high availability and flexibility of inference. It should also be noted that the correct preparation of the required information and a sound approach to the analysis process will produce results that are close to objective. Several qualitative methods are described in the literature that can be used to analyse predictive adverse events. Also, ISO 31010:2019 [11] proposes a set of qualitative methods that can be used for risk assessment. The Author recommends using the proposed risk assessment method for predictive risk analysis. However, it should be noted that qualitative methods can also be used in Model 1 if the company does not have the right tools and professionals with the skills to perform quantitative risk analysis. Examples of qualitative methods are shown in Fig. 5.15.

Fig. 5.15 Qualitative methods recommended for risk analysis in Model 2 (based on [11])

QUALITATIVE METHODS

- What-If analysis
- root cause analysis
- hazard and operational capability analysis (HAZOP)
- failure cause and effect analysis (FMEA)
- cause-and-effect analysis
- environmental risk analysis
- Bow-Tie model
- brainstorming
- benchmarking

It is also worth noting that the use of qualitative methods brings certain benefits [12]:

- There is no need to value information (availability, confidentiality, integrity).
- There is no need to quantify the impact and frequency of the risks involved.
- Estimating the costs of recommended risk treatments and calculating potential losses is unnecessary.
- Can be used without specific information and quantitative data or resources needed with quantitative methods.
- Ability to include factors in the analysis that are not quantitative, e.g., company image and organisational culture.
- Identification of areas in the organisation (process) that require special attention from the point of view of process continuity, effectiveness, efficiency and safety.

In the case of qualitative methods, the experience and knowledge of the assessment team members are of particular importance. For this reason, a company deciding on this risk assessment method should make every effort to ensure that the assessment team consists of people with extensive professional experience and interdisciplinary knowledge of the CHT system under study.

Model 3: Scenario analysis for unknown adverse events

Identifying adverse events can identify risks of an unknown nature for which the organisation has the limited knowledge to assess the probability and consequences of occurrence. This group usually includes events with a very low probability of occurring but whose consequences could even be catastrophic for the company. Precisely because of the possibility of a very high loss or risk to the health or life of process participants, this group of adverse events should also be analysed, even though the acquisition of knowledge about their course and final results is limited.

Because these risks are unknown, the number of possible methods to be used is limited. The low incidence of these adverse events makes it difficult not only to determine their impact but also to predict their course, including, in particular, the identification of risk-activating factors. Due to the limited knowledge of the phenomenon under investigation, the only risk analysis method is brainstorming with experts. The expert team should consist of external specialists and in-house staff with extensive experience and comprehensive knowledge of internal processes, but also of market trends and other external factors affecting the CHT system and the process it supports.

A tool that can further support the evaluation process is scenario analysis. The main objective of the procedure carried out is the development of scenarios, the preparation of which is intended to stimulate thinking about [13]:

- possible adverse events,
- assumptions about the occurrence of adverse events, including possible triggers for the risks involved,
- course of adverse events and the possible consequences of their occurrence.

The analysis involves the preparation of multivariate, narrative scenarios depicting the possible course and potential consequences of the assessed adverse event. Each scenario describes the phenomena caused by the occurrence of the analysed event, taking into account various internal and external factors that may affect the consequences. The correct execution of the scenario analysis is critical, as based on the results obtained (including a picture of the course of the adverse event), the organisation will implement protective barriers to reduce the consequences of the assessed risk. However, it should be noted that, as recommended in [14], the scenario analysis should also consider situations in which the controls would fail or a risk with a combination of parameters that have never been recorded before would occur. This forces team members to go beyond the control zone and take a broader view of the phenomenon under analysis when preparing the scenario analysis.

5.2.4 Risk Evaluation

The final stage of the risk assessment process is risk evaluation, which prioritises the estimated risk indicators that will form the basis for decisions on how to proceed in the risk management process. The evaluation process results will thus influence managers' attitudes towards the risks involved. Therefore, the evaluation aims to determine the rules for deciding individual risk acceptance levels, which will determine the organisation's implementation of selected risk mitigation measures and prioritise their activation.

The most straightforward rule of risk evaluation is distinguishing two basic risk classes—(a) acceptable risk and (b) unacceptable risk. However, with such division rules, it is difficult to prioritise the implemented risk mitigation measures. For this reason, most researchers introduce an additional diversification of levels for the unacceptable risk group to prioritise and rank the actions taken with limited financial

Fig. 5.16 Risk acceptance levels

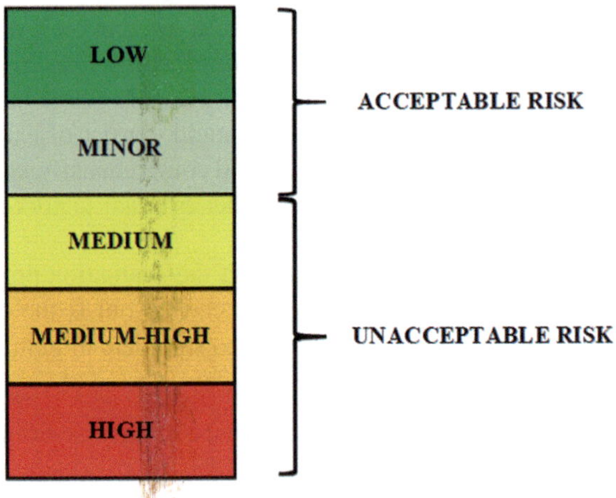

resources. This approach was also applied to the proposed risk assessment method, and the differentiated risk acceptance levels are shown in Fig. 5.16.

In accepting the risk involved, managers do not take action to reduce the risk, as in most situations, the expenditure incurred would not be economically justified. Therefore, the most common recommendation concerning this group of events is to continuously monitor the level of risk and the factors responsible for activating the threat. However, if the risk is not accepted, managers must develop a specific way of dealing with the risk. However, due to the usually limited financial resources present, it is recommended that measures be taken first to mitigate critical risks (high level). In this case, action to reduce the level of risk should be taken immediately (without undue delay—as soon as possible) and without regard to the costs involved. In the second instance, managers should take action targeting events whose risk indicators have been assessed as medium–high. The countermeasures concerning the available financial resources should be optimised in this case. Their implementation time may also be longer than for critical risk events. In the case of adverse events with risk indicators reaching medium and minor values, mitigating actions are reviewed regarding the costs incurred to achieve the expected results and are only implemented where economically justified.

Limits for individual risk acceptance levels are set on an individual basis by the risk assessment team. Decisions in this area are strongly influenced by the so-called risk appetite of managers, which is characterised in Chap. 2. Risk appetite determines managers' tolerance level to the risks involved in the planning and execution process of internal transport. This appetite thus shapes the acceptable limits of the risks involved. Therefore, it makes sense for the assessment team to include representatives of different risk appetite levels, as this will ensure that the limits set are a compromise between extreme attitudes and the risks involved.

Basing the risk index on the product of two parameters (the frequency and the consequences of the occurrence of a given event) makes it possible to use a well-known tool for risk evaluation, namely a risk matrix (map). This tool is popular

CERTAIN (P5)	MEDIUM HIGH	MEDIUM HIGH	MEDIUM HIGH	HIGH	HIGH
VERY LIKELY (P4)	MEDIUM	MEDIUM	MEDIUM HIGH	MEDIUM HIGH	HIGH
POSSIBLE (P3)	MINOR	MINOR	MEDIUM	MEDIUM HIGH	MEDIUM HIGH
UNLIKELY (P2)	LOW	MINOR	MINOR	MEDIUM	MEDIUM HIGH
RARE (P1)	LOW	LOW	MINOR	MEDIUM	MEDIUM
	NEGLIGIBLE (C1)	MINOR (C2)	MODERATE (C3)	HIGH (C4)	CATASTROPHIC (C5)

Fig. 5.17 Risk matrix

in industrial risk analysis, as it clearly and visibly positions each of the assessed adverse events and allows them to be prioritised for risk mitigation. Thus, it provides a map for managers to set directions for dealing with risks. For the distinguished risk acceptance levels shown in Fig. 5.16, a matrix was developed to evaluate the risk parameters on a scale of 1–5. This map is presented in Fig. 5.17. However, it should be remembered that the limits of the division are left to the discretion of the decision-makers, and those presented in the matrix are only proposals made by the Author.

However, it should be remembered that the limits of the division are left to the discretion of the decision-makers, and those presented in the matrix are only proposals made by the Author.

Based on the evaluation, managers decide on the recommended attitudes towards the listed risk groups. These attitudes will depend primarily on the resources available to the organisation and the maturity of the company's risk management system. It should be taken into account that the chosen risk management system dictates how the management process will proceed, and this will determine the process's resilience to adverse events.

5.2.5 Risk Monitoring

Risk assessment and monitoring must be continuous in the organisation. Relying on sporadic risk analyses initiated after an adverse event is an unnecessary activity that inefficiently engages resources. It therefore makes sense to implement a risk culture in the company based on:

- the periodic, regular process of updating the risk assessment for the internal transport process and
- the constant monitoring of the activity of the CHT system elements and their collaboration affects the correct execution of the internal transport process and the results achieved.

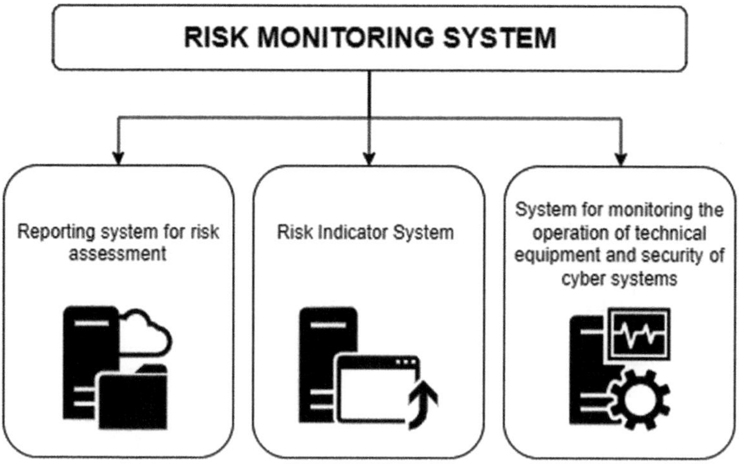

Fig. 5.18 Components of a risk monitoring system

When describing the risk identification and analysis phase, particular attention was paid to the availability of complete and up-to-date data. The data quality feeding the risk assessment process depends mainly on the risk monitoring systems in place. The risk monitoring process associated with the operation of CHT systems should be based on the three components shown in Fig. 5.18.

(1) **Reporting system for risk assessment**

A fundamental challenge in preparing a risk assessment is the limited availability of comprehensive data on adverse events occurring. The lack of such data is often due to inadequate procedures indicating the extent of the information needed and the means of collecting it, including the persons responsible for the process. Such procedures only take shape as the risk culture develops in the organisation, and information technology can strongly support the process. For the process to be effective and efficient, it is necessary to develop assumptions for a reporting system for risk assessment.

In the proposed risk assessment method, the assessment team focuses on the course (execution) of the internal transport process and the results achieved by the service process. Therefore, the reporting system should be based on the two pillars shown in Fig. 5.19:

- a knowledge base of occurring adverse events—collecting comprehensive information on disruptions and adverse events affecting the course of cargo handling in internal logistics and
- a performance database that provides a basis for analysing potential deviations from service targets/standards

The adverse event report is designed to collect detailed data on recorded adverse events. In the document template, the responsible persons should enter the required

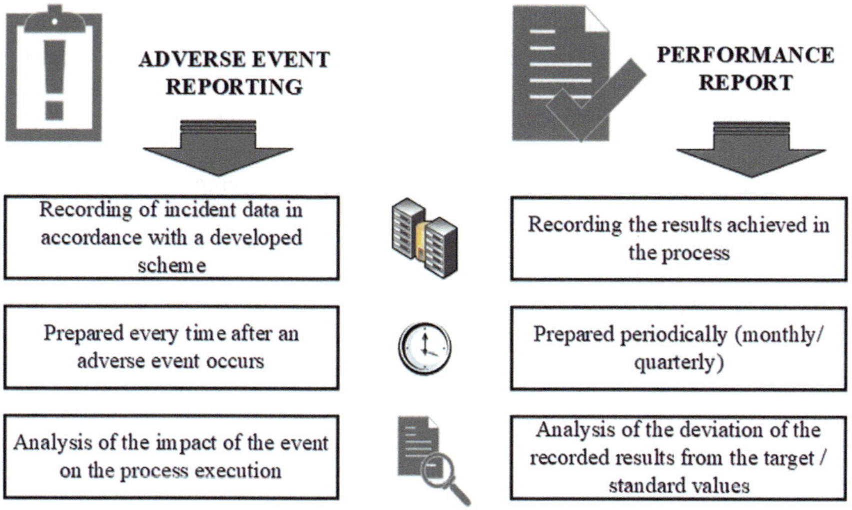

Fig. 5.19 Pillars of the reporting system for risk assessment

data. Wherever possible, suggested answers should appear under these headings, and only in justified cases should the recorded data be left free. In this way, the data to be entered will be standardised, significantly speeding up the procedure for further analysis. The duty to record data on an adverse event lies with the employee responsible for monitoring the execution of the process. However, in the case of the CHT system, some of the data may come from selected IT systems responsible for supervising the correct operation of technical equipment and monitoring its operational parameters. The range of data reported should meet the information needs of the assessment team, but among the primary data that need to be recorded are:

- date of the adverse event;
- the type of event (short characteristics);
- the cause of occurrence (concerning the analytical layers);
- the consequences of the adverse event;
- the environmental conditions at the time of the event;
- the recovery time of the CHT system after the adverse event.

This report data will provide decision-makers access to complete and up-to-date knowledge of adverse events. It is also worth noting that the available Industry 4.0 solutions implemented within CHT systems can significantly automate the process of collecting and analysing input data. An appropriately defined framework for the recording system of adverse event information can improve the effectiveness, efficiency, and time required to make inferences for risk assessment purposes. The knowledge base created from the reported data also provides information for determining selected Risk Indicators.

The performance report aims to support monitoring the correct execution of the internal transport process by monitoring the results achieved, expressed as Key Performance Indicators (KPIs). KPIs can be quantitative indicators (e.g. lead time, logistics service level, number of vehicles involved) or value indicators (e.g. energy costs, employee salary costs). Based on the reported data on current KPIs, deviations from set targets (defined standard values) are analysed. The template for the reported data and the indication of the source are presented in Table 5.2.

This report should be prepared periodically—ideally once a month, but at least once a quarter. Excessive intervals between reports may limit the flexibility of managers' responses to disruptions as they occur.

(2) **Key risk indicators system**

The reporting system should be integrated with a system of indicators, the scope of which should take into account the information needs of the risk assessment team. The particular attention of decision-makers is primarily focused on performance measures (concerning the effectiveness and efficiency of the activities undertaken) and the demand for resources to perform assigned tasks. Therefore, the measurement system developed should be based on traditional performance indicators, including task performance and resource utilisation indicators, as well as risk indicators, which are calculated based on performance indicators and data from adverse event reports. In this case, the risk indicators have an informative role (providing data for the risk assessment to be carried out) and should also act as gatekeepers to correctly executing processes. In the case of changes in the values of risk indicators indicating an increase in risk, it is necessary to take appropriate action, including carrying out proceedings to assess the impact of the change on the level of risk associated with the analysed process.

The selection of indicators forming the set of performance and risk indicators is an individual decision of each enterprise. Managers should make this decision based on a thorough analysis of the information needs of process owners and the risk assessment team. However, it should be remembered that too many indicators that require ongoing monitoring increase the operational costs of the assessment process. Therefore, it is reasonable to distinguish from the set of performance and risk indicators a group of critical indicators called Key Performance Indicators (KPIs) and Key Risk Indicators (KRIs), whose ongoing monitoring managers consider critical

Table 5.2 Model performance report

Background information	Method of registration
Indicator name	Downloading from information systems
Target indicator value	Downloading from information systems
Current indicator value	Downloading from cyber-systems
Recorded deviation	Automatic calculation
Identified causes of deviation	Input by the manager/downloading from cyber-systems
Identified effects of deviation	Input by the manager

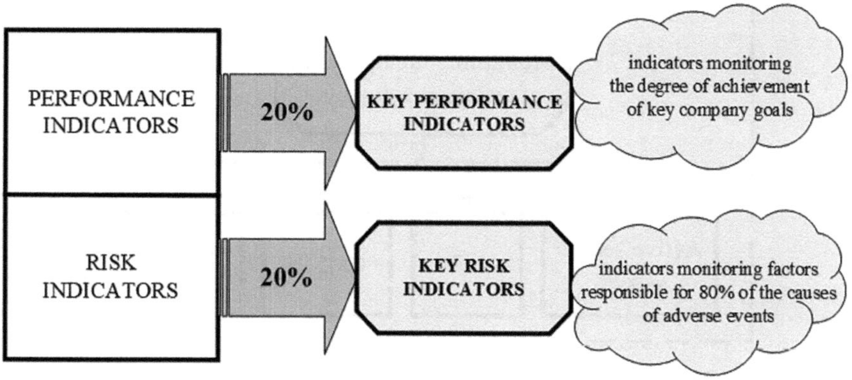

Fig. 5.20 Pareto rule in the determination of KPIs and KRIs

from the point of view of enterprise risk management. The Pareto method (20/80 rule) can distinguish between KPIs and KRIs, allowing decision-makers to identify, for example, 20% of the indicators monitoring the factors responsible for 80% of the causes of adverse events. This rule is illustrated in Fig. 5.20.

For the group of KPIs and KRIs, it is recommended that managers select indicators that are used operationally as part of their day-to-day monitoring of process correctness or are considered important from the perspective of improving the assessed process. It should also be remembered that the group of KPIs should include at least one representative for each analytical layer highlighted.

(3) **Technical equipment performance monitoring and security cyber systems**

The essential components of the risk monitoring process are the technical equipment performance monitoring systems and security cyber systems, shown in Fig. 5.21

CHT systems' support of the transport process offers a vast potential for monitoring the operation of the technical equipment used. Industry 4.0 solutions based on collections of active sensors monitoring selected elements and activities of technical equipment and the Internet of Things make it possible to track in real time a broad spectrum of parameters related to the operation of a given piece of equipment and deliver them to a central system collecting and analysing the measurement results (CMMS—Computerised Maintenance Management System). Based on defined limit values, these systems not only report on deviations from the accepted standards but can also recommend actions or initiate reactions of other digital systems according to a procedure algorithm implemented for these purposes. The basic functionality of CMMS systems includes:

- asset management (asset tracking, recording asset history);
- scheduling labour and resources;
- work order management;
- preventive maintenance plan;
- stock management;

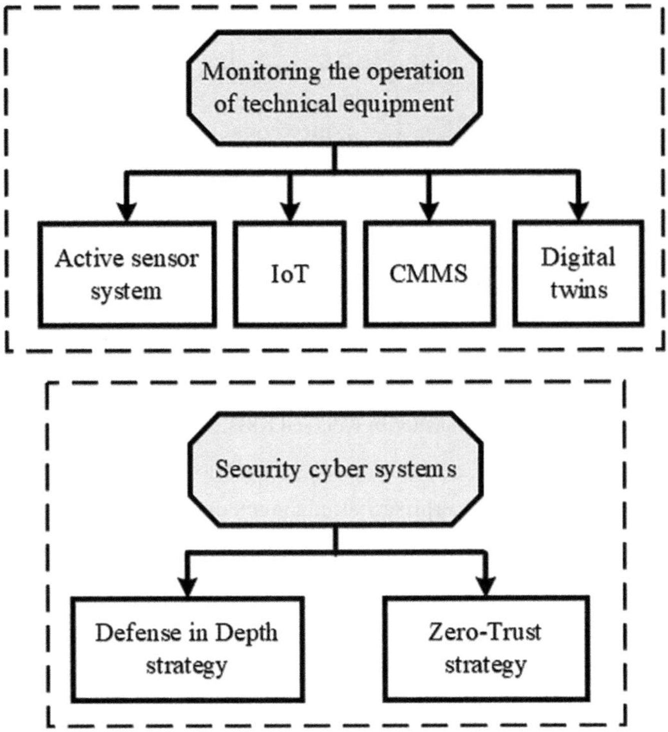

Fig. 5.21 Technical equipment performance monitoring and security cyber systems for operational risk assessment

- audits;
- performance indicators and reports.

However, in CHT systems, in addition to CMMS systems and collections of active sensors, digital twins, the idea of which is described in Chap. 3, are an important solution to support the monitoring process. Digital twins can be created to digitally represent the vehicles used in the material handling process. By exchanging data in real time between the physical object and the digital object, it is possible to:

- monitor the operation of technical objects in real-time;
- help to understand better disruptions occurring in the operation of technical facilities;
- accelerate the response to disturbances occurring in the operation of technical facilities;
- anticipate the occurrence of disturbances (failures), indicating the type and time of their occurrence;
- analysing the response of a technical facility to planned actions involving it;
- elimination of erroneous activities and inefficient investments.

As has been highlighted on many occasions, CHT systems are particularly vulnerable to digital attacks, which can come from within the organisation as well as from the environment. Therefore, systems that monitor the security of cyber systems are a critical part of the risk monitoring process. The security monitor should be concerned with the data collected, the network, the entire IT infrastructure, and the software and mobile devices. This process should be supported by appropriate cyber security systems, such as anti-virus software, firewalls, identity and access management systems (allowing you to determine who has access to your organisation's resources and who can change them). Two of the most popular strategies for securing industrial networks today include the concepts of Defense in Depth and Zero-Trust.

Defense in Depth uses a combination of different defence mechanisms to protect an organisation's critical resources from potential attack [15]. Therefore, this strategy creates a multi-layered security system to mitigate risk and minimise potential losses. Creating a multi-layered security architecture provides multiple lines of defence against cyber threats. In this way, despite the breach of one security layer, other layers remain unaffected, as each layer represents a different security control or mechanism. Comprehensive guidance on the security layers to be created can be found in the literature and subject portals. Abdelghani [16] emphasises that the Defense in Depth strategy is based on technology, people and processes. Therefore, among its basic components, there are[16]:

- Physical Access Protection.
- A comprehensive set of policies and procedures covering all cybersecurity aspects.
- Network Segmentation and Separation—physical separation, network segmentation, Firewall and VPN.
- Device hardening operations, which include: (a) password management including encryption, (b) disabling of unused services, (c) access control, (d) patches, hotfixes, application updates, and (e) strong authentication.
- Monitoring and maintenance.

The second Zero-Trust strategy is based on the premise of "trust no one, verify everything". Earlier strategies considered a perimeter-based security model (every user, device, and network traffic inside the border is trusted). In contrast, the zero-trust model treats everything as hostile, and trust must be gained through multiple parameters, including user authentication, authorisation, and verification of devices and services. Among the main principles of this strategy are [17]:

- All critical data, assets, and services must be secured irrespective of location (inside or outside the corporate network).
- No components are trusted by default, and access communications must always be encrypted, even for intranet channels.
- Just-in-time access, where the authentication and authorisation verification are done based on the set access policy precisely at the time of access request and will only be valid for one session.

- Just enough access, which corresponds to least privilege access, should be applied such that only enough amount of access required to carry out the task is provided and no more.
- Access control policies must be configured based on the data received from a maximum number of sources, such as device health, type of resource accessed, etc.
- Access is never granted based on history and must be evaluated whenever requested.

At the same time, it is crucial to remember that the Zero-Trust strategy is a concept that uses various security solutions that target the elimination of potential vulnerabilities present in the security layers. Critical to this is the fact that this model covers not only traditional digital assets but also addresses the protection of the cloud space, IoT, and other Industry 4.0 solutions implemented in smart factories [17].

5.3 Implementation of the Proposed Method ORA_CHTS in Internal Transport Systems

The framework of the proposed risk assessment method is universal and can be implemented in various internal logistics processes supported by CHT systems. However, the object of the research described in this book is the internal transport process. Therefore, the applicability of the proposed ORA_CHTS method in the assessment procedure will be presented using this process as an example.

5.3.1 Characteristics of the AGV System

The ORA_CHTS method was implemented in a specific internal transport system handling material flows between the warehouse and the production floor in a selected company from the automotive sector. Transport handling concerns delivering parts from the material warehouse to the assembly lines, picking up containers with finished products from the assembly stations, and transporting them to the warehouse. The assembly stations were handled in 450 m and 350 m loops, which start and end at the parking point. The layout of both routes is shown in Fig. 5.22.

Until recently, the material handling process was executed by vehicles manually operated by skilled operators. This system was replaced by an AGV solution using laser navigation. The choice of navigation method was mainly influenced by two factors: (a) the implementation of the AGV system into the existing technical and construction infrastructure and (b) the resulting contamination during metal processing in the vehicles' moving space. The significant benefits of laser navigation are primarily the flexibility of planning the paths of the vehicles, ensuring that periodic changes can be made. The vehicle is equipped with a navigation module

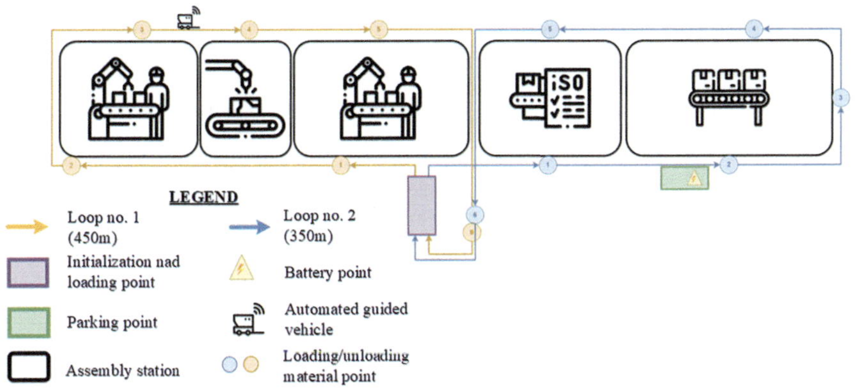

Fig. 5.22 Layout of the transport system work area

containing a laser that is responsible for reading information from reference markings placed on the hall equipment. The laser sensor has an optoelectronic detection system, a visualisation LED, a 7-segment digital readout and an optical shield. The other parameters of the AGV system used are shown in Table 5.3.

The vehicle has a touchscreen graphic interface allows communication between the vehicle and a human. The most important messages are shown in Fig. 5.23.

Prior to the start of a mission, the operator is required to indicate in the AGV system the stations that have been scheduled for service. The scope of the mission being carried out is displayed on the screen, indicating the unloading and loading locations. When the mission is completed, the vehicle returns independently to the charging point and waits for the next mission to start. When the vehicle's battery charge level reaches a critical level, the vehicle returns to the parking area, which also acts as a battery charging point. The need to change the battery is reported on the truck's control screen, and, in addition, light and sound signals are given to alert the operator. A new mission can only be started once the batteries have been replaced.

In addition, the vehicle is equipped with dedicated measures to ensure its safe operation due to sharing the workspace with people:

- The stacking control module consists of four controls with different colours that activate in specific situations. Table 5.4 shows the meaning of the activated controls.

- Emergency stop button—its use activates the brake and cuts off traction: all moving parts of the vehicle lose power. Resumption of vehicle operation is only possible after operator intervention.
- The dynamic deceleration field—a virtual field created in front of the tractor at a height of 170 mm, which activates when an obstacle is detected in this zone. In such a situation, the vehicle reduces its speed to 0.72 km/h. The field is dynamic, as the programming adjusts the shape of the field to follow the actual path.

Table 5.3 Parameters of the AGV system

Communication module	WiFi
Maximum vehicle speed	1.38 m/s = 4.97 km/h
Deviation from a given position in free movement	± 50 mm
Number of parking spaces	1
Required clearance on both sides of the robot in a parking space	50 cm
Number of charging points	1
Low battery	30%
Battery charge critical threshold	20%
Access to the charging point	Manual
Floor requirements	Flat, stable, dry—deviations from specifications should be assessed locally
Temperature	0–40 °C
Maximum temperature	40 °C should not last longer than 60 min
Average temperature	up to 35° in 24 h
Temperature change	up to 5 °C per hour
Humidity range	45–90% przy 20 °C
Maximum slope on the route	3%
Minimum height on the route	2.5 m
Maximum load capacity	5 t
Maximum overall dimensions of the load	2.63 m × 1.063 m
The minimum distance of an object from the ground necessary to be detected by a safety laser	10 cm

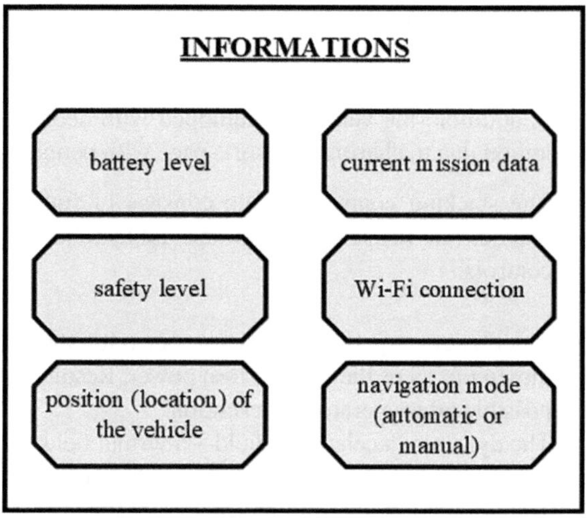

Fig. 5.23 Information presented on the graphic interface

Table 5.4 Stacking control module

Controlled by the module	Mode			Meaning
Module 1	constantly on			detained for safety reasons
	flashing			waiting for the operator to restart
	off			safety OK
Module 2	constantly on			system error
	flashing			low battery
	off			system OK
Module 3	constantly on			manual mode stationary robot
	flashing			manual mode robot in motion
	off			robot inactive
Module 4	constantly on			automatic mode stationary robot
	flashing			automatic mode robot in motion
	off			robot inactive

- Dynamic stopping field—a virtual field created in front of the tractor at a height of 170 mm, which activates when an obstacle is detected in this zone. In this situation, the vehicle comes to an emergency stop. This field is also dynamic and is located in front of the slowing field, as shown in Fig. 5.24.

- Laser curtain scanner for obstacle detection—in the event of an obstacle appearing in the area of the laser curtain's designated field, the vehicle stops, simultaneously sending out an audible and luminous warning signal.
- Blue spot in front of the vehicle—a blue LED light informs employees of an approaching vehicle. This light is significant at intersections of vehicle and people paths, so the point is 4 m away from the vehicle.
- Traffic lights for correct train connection (tractor and wagons)—a traffic light is placed on each wagon and can send a green or red light signal. The green part of the traffic light indicates that all train elements are correctly connected. A red light indicates that there is some irregularity in the formation.

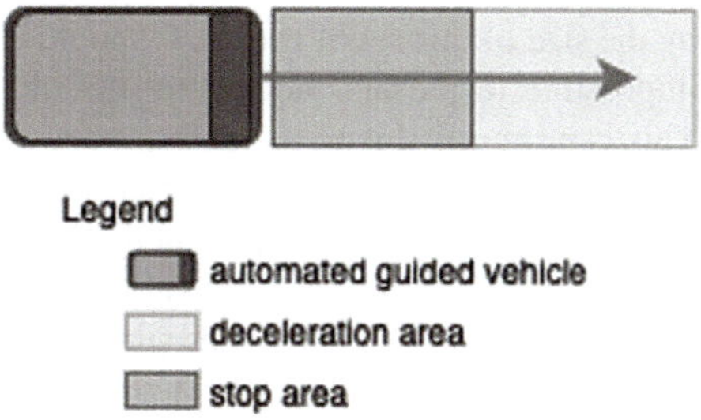

Fig. 5.24 Dynamic slowing and stopping field

It is also worth mentioning that if the vehicle loses its reference point to the mapped path and the navigation laser cannot measure enough points, it immediately enters error mode.

5.3.2 Risk Assessment for a Selected Transport System

An interdisciplinary team of experts was appointed to prepare the risk assessment. The team consisted of (a) the director of the logistics division, (b) the warehouse manager, (c) the production director, (d) the AGV system operator, (e) the leader of the cargo handling team, (f) a scientist—a specialist in risk management in logistics systems (the Author). The expert team used the ORA_CHTS method developed by the Author to prepare the risk assessment.

Stage 1: Setting the context for the risk assessment

The risk assessment concerned the implemented AGV system for handling material flows between the warehouse and the production floor, as described above. The adopted context for the risk assessment was to evaluate the risks by the planning and execution phases of the material handling process, using the team's expert knowledge and initial experience gained during the testing phase of the AGV system. The identification of adverse events was based on the four analytical layers identified (human factor, technical, digital and environment), considering the sources of adverse events. Accordingly, the risk identification process was carried out according to the flowchart in Fig. 5.8.

Stage 2: Identification of adverse events

The procedure for identifying adverse events related to the planning and execution of the internal transport process handled by the CHT system identified 84 adverse events. The most challenging part was the identification of unknown risks—the team identified only seven such events: 3 for the planning phase (including 1 in the human layer and 2 in the environment layer) and 4 in the execution phase (including 1 in the technical layer and 3 in the digital layer). The risks identified and their classification into the distinguished classes are shown in Table 5.5.

The low number of identified events for unknown risks may be because most of the identified adverse events were categorised as known risks, which is also confirmed by the size of this set of events. It should also be emphasised that it is unfortunately impossible to present the full results of the known risk assessment in the book. This is primarily due to the extensive research material covering the analysis of 84 events and the company's protected know-how. However, to verify the usefulness of the developed ORA_CHTS method, two exemplary adverse events belonging to each distinguished class of the known risk category were selected. These events are presented in Table 5.6, together with their respective classifications.

For the adverse events indicated in Table 5.6, a risk analysis was performed using Model 2.

Table 5.5 Unknown risks identified for the assessed system

Name of the adverse event	Class affiliation
Delegating the control of the AGV system to an inexperienced operator	AE-H/PP/UK
Bankruptcy of the AGV system service provider	AE-E/PP/UK
Introduction of high limits on the consumption of energy from non-renewable sources	AE-E/PP/UK
Failure of the AGV system due to operation in very high temperatures (occurrence of hot weather and limited use of air conditioning)	AE-T/EP/UK
Incorrectly prepared update of the AGV system control software	AE-C/EP/UK
Internal cyber-attack on the AGV system	AE-C/EP/UK
Uncontrolled activation of the fire extinguishing system in the server room, leading to damage to the servers	AE-C/EP/UK

Table 5.6 Known risks identified for the AGV system under assessment

Event symbol	Name of the adverse event	Class affiliation
AE_C_1	Failure to implement an adequate digital security system	AE-C/PP/K
AE_C_2	Incorrectly parameterised human-vehicle communication system	
AE_C_3	External cyber attack	AE-C/EP/K
AE_C_4	Temporary interruption of energy supply	
AE_H_1	Incorrect assignment of missions to vehicles	AE-H/PP/K
AE_H_2	Incorrect assignment of paths	
AE_H_3	Human intrusion into the path of a vehicle	AE-H/EP/K
AE_H_4	Failure of an employee to see a moving vehicle	
AE_T_1	Incorrectly mounted laser sensor on AGV	AE-T/PP/K
AE_T_2	Incorrectly programmed warning (limit) parameters for vehicle operation	
AE_T_3	Collision between vehicle and another vehicle	AE-T/EP/K
AE_T_4	Excessive speed of the vehicle in the shared working space	
AE_E_1	Incorrectly prepared surface on which vehicles are running	AE-E/PP/K
AE_E_2	Incorrect marking of the area not accessible for vehicles	
AE_E_3	Incorrect lighting interfering with the reading of the reference mark	AE-E/EP/K
AE_E_4	Pollution of the hall—oil spills, production waste	

Stage 3: Risk analysis

The company does not yet have a sufficiently complete database, as the AGV system has only just been implemented for cargo transport. This makes it impossible to perform a risk assessment using quantitative methods to estimate the frequency parameters and consequences of identified adverse events. For this reason, it was not possible to use Model 1 to prepare the risk analysis. However, in the future, once the company has collected adequate quantitative data describing the operation of the AGV system, it is planned to implement Model 1 into the risk assessment process.

A classic brainstorming technique was used with no time limit and a defined number of ideas to analyse events belonging to the set of unknown risks. This provided the assessment team with the freedom to proceed with the analysis. The analysis was primarily a cause-and-effect analysis. The team first identified the potential sources of occurrence of each of the seven identified adverse events, and then the possible consequences of their occurrence were determined. For this part of the risk analysis, the qualitative Bow-Tie method and scenario analysis were used to develop multivariate, narrative scenarios describing the possible course and potential consequences of the assessed adverse event. Unfortunately, these research results cannot be published due to the company's safety policy and know-how. For this reason, in the remainder of the book, the presentation of the results is limited to events belonging to the known risk category.

Due to the lack of historical data, the assessment team used Model 2 using fuzzy set theory, described in Sect. 5.2, and the Mamdani model shown in Fig. 5.14 to assess the known risk. The team used the Mamdani model, built in Matlab software, to prepare the risk analysis. In the model, the input variables were the probability and consequences of the identified adverse events, and the output variable was the event's risk level. The probability and consequences of each adverse event were estimated based on the expert opinion of the assessment team members. Linguistic variables were used to express opinions on the two parameters assessed, which were then modelled using fuzzy set theory. Table 5.7 (probability) and Table 5.8 (effects) show the linguistic variables for the input variables. The linguistic variables for the output variable (risk level) are presented in Table 5.9.

As described in Sect. 5.2, the risk parameters of each analysed adverse event are treated as fuzzy numbers (FN) and membership functions (MF). A triangular MF (for the probability and consequences of the adverse event) and a trapezoidal MF

Table 5.7 Probability of adverse event

Rating category	Description	Fuzzy value
CERTAIN (P5)	Expected to occur regularly under normal circumstances	(7.5, 9, 10, 10)
VERY LIKELY (P4)	Expected to occur at some time	(5.5, 6.5, 8)
POSSIBLE (P3)	May occur at some time	(3.5, 4.5, 6)
UNLIKELY (P2)	Not likely to occur in normal circumstances	(1.5, 2.5, 4)
RARE (P1)	Could happen, but probably never will	(0, 0, 0.5, 2)

Table 5.8 Consequences of adverse event

Rating category	Description	Fuzzy value
CATASTROPHIC (C5)	Failure to complete the mission entirely, permanent damage to the AGV, loss of the order/client, permanent damage to workers' health	(7.5, 9, 10, 10)
HIGH (C4)	Failure to complete the mission, crash or damage to the AGV, injured worker	(5, 6.5, 8)
MODERATE (C3)	Inability to start the mission, stopping the operation of the AGV system, need to postpone the service to a later date, risks to workers' health and safety	(3, 4.5, 6)
MINOR (C2)	Disruptions cause a delay in the implementation of the service or the lack of the expected results of the mission (in terms of the quality and completeness of the measurements performed)	(1, 2.5, 4)
NEGLIGIBLE (C1)	No impact or little impact on security	(0, 0, 0.5, 1.5)

Table 5.9 Risk level

Rating category	Description	Fuzzy value
HIGH (RL5)	Consequences are catastrophic or high for AGV missions and the service provided, which occur certainly or almost certainly in the future	(80, 90, 100 100)
MEDIUM–HIGH (RL4)	A certain (almost certain) event with minor consequences or an unlikely event with critical consequences	(60, 70, 80, 90)
MEDIUM (RL3)	An almost certain event with minor consequences or an unlikely event with significant consequences. Limiting the possibility of a successful AGV mission	(40, 50, 60, 70)
MINOR (RL2)	The occurrence of an event is possible—minor/no significant consequences without much impact on the mission's success	(20, 30, 40, 50)
LOW (RL1)	It is almost impossible for an event to occur. The consequences are negligible—no impact on the accomplishment of the mission	(0, 0, 15, 25)

(for the level of risk) were chosen for the analytical procedure. This selection was based on the popularity of both functions and the best fit to the data. The arithmetic mean aggregation operator was used to aggregate the collected opinions of the team members. The aggregation represent the operator, defined on triangular FN: (a_1, b_1, c_1), $(a_2, b_2, c_2) \ldots (a_n, b_n, c_n)$ delivers the result as (x, y, z) according to the formulae 5.13:

$$\begin{cases} x = \frac{1}{n}\sum\limits_{k=0}^{n} a_k \\ y = \frac{1}{n}\sum\limits_{k=0}^{n} b_k \\ z = \frac{1}{n}\sum\limits_{k=0}^{n} c_k \end{cases} \tag{5.13}$$

where n is the number of experts (in the analysis, $n = 6$).

The membership functions (triangular and trapezoidal) used in the model to represent the linguistic scales of the input and output parameters were constructed from Tables 5.7, 5.8 and 5.9. They are shown in Figs. 5.25 and 5.26.

The surface risk results are visualised in Fig. 5.27 for the risk assessment model built in Matlab. It should be noted that no rule prioritisation was applied in the model, which means that all entered rules in the model have a weight of 1.

The result of the risk analysis carried out using the Mamdani model is shown in Table 5.10.

The results clearly show a wide variation in the risk levels estimated for the individual adverse events. The highest risk level concerned event AE_C_4 (Temporary interruption of energy supply) and AE_C_3 (External cyber-attack), while AE_E_1 (Excessive speed of the vehicle in the shared working space) reached the lowest risk level. The values of the risk indicators for the analysed adverse events form the basis for further evaluation of the results obtained.

Stage 4: Risk evaluation

A risk matrix was used to evaluate the results obtained. The matrix was built based on the rule base entered into the Mamdani model and the defined risk acceptance levels shown in Fig. 5.16. The matrix denotes classifying individual adverse events into the corresponding risk categories. This classification is shown in Fig. 5.28.

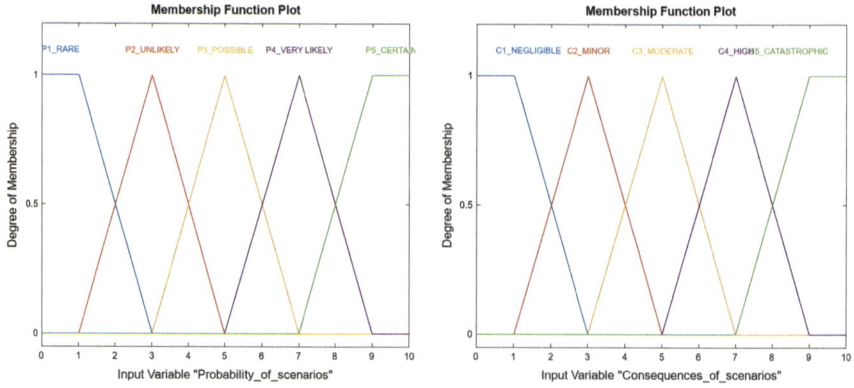

Fig. 5.25 Membership functions of probability and consequences

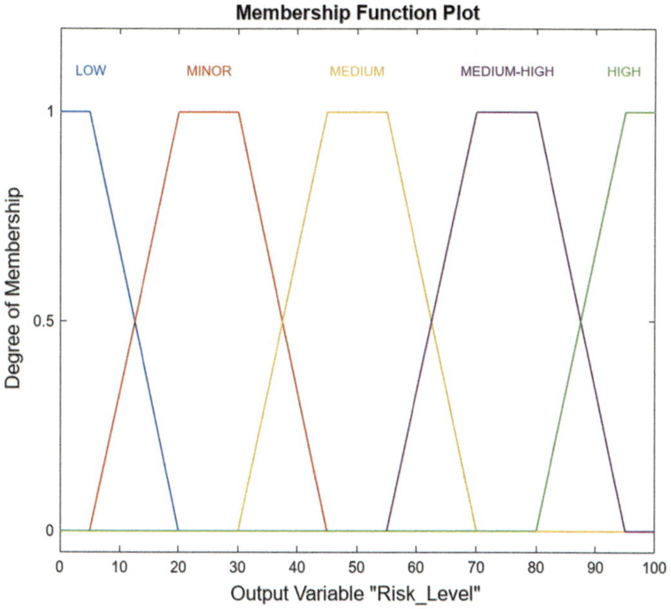

Fig. 5.26 Membership functions of risk level

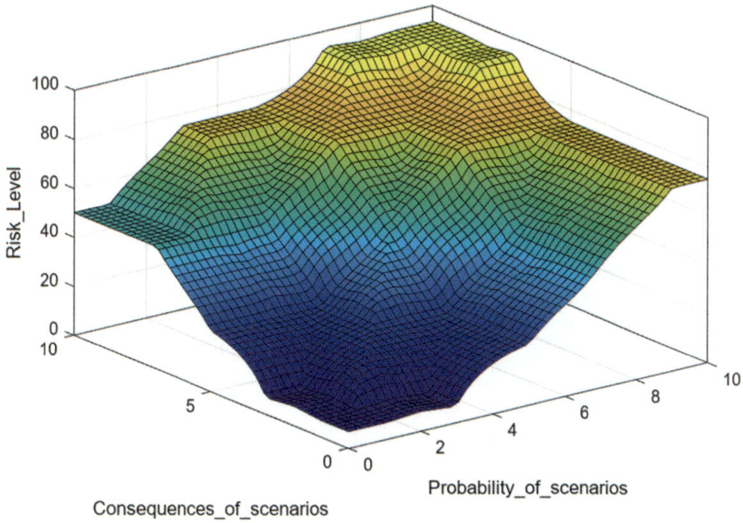

Fig. 5.27 Surface risk results

Table 5.10 Results of risk analysis for adverse events in the CHT system

Event	P (Fuzzy)	C (Fuzzy)	P (Crisp)	C (Crisp)	Risk level
AE_C_1	P2	C4	3.33	7.12	54.81
AE_C_2	P3	C4	5.12	7.33	75.31
AE_C_3	P3	C5	4.66	8.56	70.07
AE_C_4	P4	C4	7.22	7.07	75.61
AE_H_1	P2	C3	2.89	5.23	28.40
AE_H_2	P2	C4	2.96	6.88	48.10
AE_H_3	P3	C4	5.22	6.96	74.30
AE_H_4	P3	C3	6.56	5.04	68.80
AE_T_1	P2	C4	3.56	6.66	52.50
AE_T_2	P2	C4	3.03	6.84	48.34
AE_T_3	P2	C4	3.07	6.67	46.40
AE_T_4	P2	C4	2.86	7.34	54.90
AE_E_1	P2	C3	2.95	5.08	26.33
AE_E_2	P2	C4	2.88	7.07	51.15
AE_E_3	P3	C3	4.89	4.91	48.30
AE_E_4	P3	C4	5.34	6.98	73.00

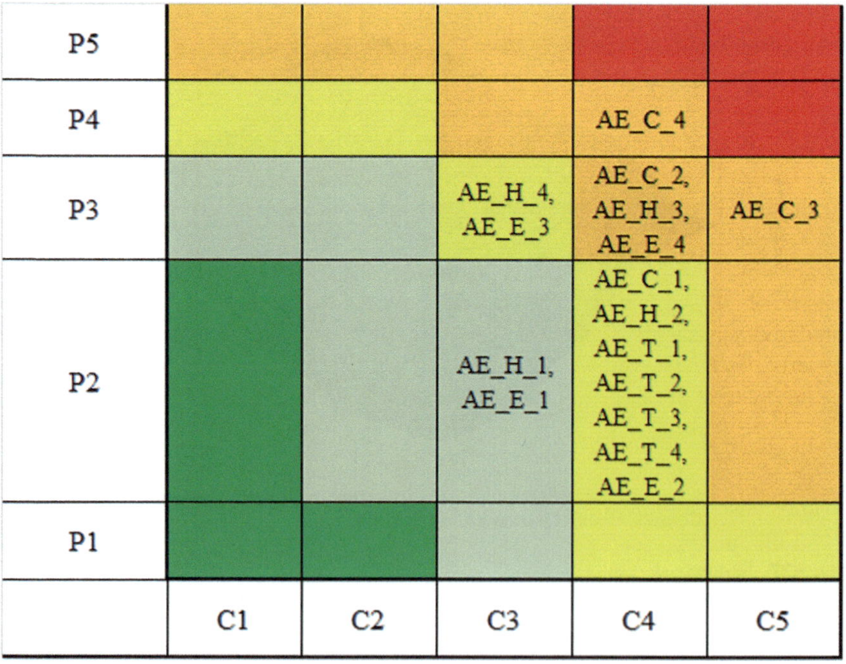

Fig. 5.28 Risk evaluation results for the CHT system

The analysis of the risk evaluation results shows that none of the analysed adverse events were classified in the lowest (LOW) or highest (HIGH) risk class. The most numerous class is the MEDIUM level risk, in that most events were characterised by probability (P2) and consequence (C4). Their risk should be considered acceptable for the two events belonging to the MINOR class. For the remaining adverse events, the company is required to take specific actions to reduce the occurrence of the identified risks.

Based on this classification, it is possible to prioritise the risk mitigation actions that need to be taken. In the first instance, mitigation actions should address events AE_C_4 and AE_C_3, followed by events AE_C_2, AE_H_3, and AE_E_4. Only after these have been completed should analyses be prepared to confirm the economic viability of implementing risk mitigation solutions for events in the MEDIUM class. In the case of acceptable risk, it is recommended only to monitor it for a possible increase in the risk parameters describing it. It is also worth noting that, as a result of the analysis, none of the events reached the HIGH level, which would require immediate measures to improve the CHT system.

Step 5: Risk monitoring

The verification method focused primarily on confirming the correctness of the risk assessment steps. The verification scope carried out this way was primarily due to the research formula adopted. The developed ORA_CHTS method was not permanently implemented in the company but only applied as an experiment to assess the risk for the implemented AGV system. Therefore, it did not include a risk monitoring phase, as this would have required a comprehensive design of the monitoring system based on the three pillars highlighted. As the method description highlights, the monitoring process is an ongoing activity. At the same time, the monitoring system for the correctness of the process and the impact of the environment has to be customised to the needs of the company and take into account the specifics of the process, the risk culture applicable and the organisation's digital maturity. For this reason, verification of step 5 as a research experiment proved not feasible. The developed risk monitoring system adapted to the needs of the ORA-CHTS method was therefore only reviewed by the team of experts participating in the study being conducted. The experts positively evaluated all three highlighted pillars of the monitoring system while pointing out that:

- the designed reporting system should include the possibility of automatic data transmission from selected sensors monitoring the operation of the CHT system;
- the construction of the reporting system should be preceded by a detailed analysis of the information needs of managers and the information potential of currently used IT systems;
- risk indicators should have the same rank (importance) in the company as performance indicators and be subject to the same supervision standards;
- the system for monitoring the technical condition of equipment and digital security should be given critical priority due to its importance in keeping the process running.

The opinions formulated by the experts during the assessment detail the guidelines described by the Author in the ORA_CHTS method and are in line with her assumed approach to the risk monitoring process.

5.3.3 Comparative Analysis of Risk Levels for HT and CHT System

For the purposes of the results presented in this book, an additional risk analysis was also conducted for the selected 16 adverse events regarding their occurrence in the process handled by the traditional HT system. The same group of experts participated in the assessment because the members of this team also have a great deal of experience and knowledge of the traditional material handling system, which previously handled the material flow process between the warehouse and the assembly hall. They estimated the risk parameters for the same process the HT (human-technical) system handled based on their knowledge and experience. The Mamdani model, which was also used for the risk analysis for the CHT system, was again used for the risk analysis. The results obtained are shown in Table 5.11.

The risk levels for the same adverse events occurring in the HT and CHT systems-supported internal transport processes are compared and shown in Fig. 5.29.

Table 5.11 Results of the risk analysis for adverse events in the HT system

Event	P (Fuzzy)	C (Fuzzy)	P (Crisp)	C (Crisp)	Risk level
AE_C_1	P3	C1	4.34	1.13	22.50
AE_C_2	P1	C2	1.08	2.88	8.68
AE_C_3	P1	C2	1.33	3.11	13.30
AE_C_4	P1	C1	1.21	0.97	6.94
AE_H_1	P2	C2	2.87	3.04	24.60
AE_H_2	P2	C2	2.67	2.69	24.01
AE_H_3	P3	C3	5.11	4.22	41.80
AE_H_4	P3	C2	5.43	3.33	37.60
AE_T_1	P1	C1	0.97	1.03	6.75
AE_T_2	P2	C2	3.05	2.92	24.80
AE_T_3	P2	C4	3.67	7.22	58.93
AE_T_4	P2	C3	3.02	5.34	30.40
AE_E_1	P2	C1	2.87	1.12	9.52
AE_E_2	P2	C1	2.56	0.96	7.21
AE_E_3	P1	C1	1.13	1.77	11.10
AE_E_4	P3	C2	5.55	3.04	33.10

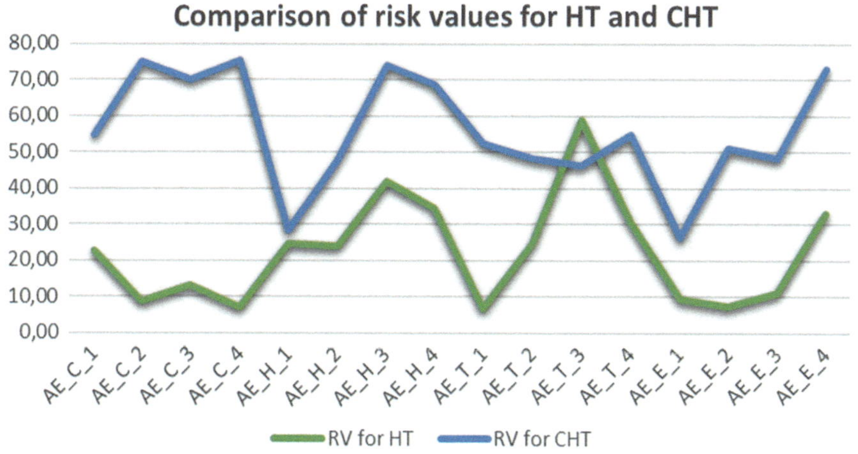

Fig. 5.29 Comparison of risk levels for the material handling process supported by the HT and CHT systems

The risk indicators estimated for the HT system were taken to the evaluation stage, in which the risk matrix was also applied. The results obtained in this stage are shown in Fig. 5.30.

As shown in Fig. 5.30, the range of classification of assessed adverse events has completely changed. Almost all the adverse events assessed have a very low-risk rating categorised as acceptable. Only two events were categorised as unacceptable risk, but this is still the MEDIUM level. It means that before taking action to

P5					
P4					
P3	AE_C_1	AE_H_4, AE_E_4	AE_H_3		
P2	AE_E_1, AE_E_2	AE_H_1, AE_H_2, AE_T_2	AE_T_4	AE_T_3	
P1	AE_C_4, AE_T_1, AE_E_3	AE_C_2, AE_C_3			
	C1	C2	C3	C4	C5

Fig. 5.30 Risk evaluation results for the HT system

reduce the risk of this event, it is necessary to analyse the economic feasibility of the implemented risk mitigation mechanisms.

The analysis of the results in Figs. 5.29 and 5.30 shows a significant discrepancy between the estimated levels of risk present in the CHT and HT system. Of course, this is understandable, as the selection of adverse events for the presented analysis results was primarily aimed at illustrating the risks associated with the operation of CHT systems. Their identification was also driven by the priorities defined in step 1, i.e. during the identification of the context of the risk assessment performed. However, the analysis of the two figures is intended to raise awareness of the need for a change in the approach to the risk assessment being carried out. Events whose probability and impact in HT systems were estimated at a low level and were monitored on an occasional basis so that no precautionary action was taken against them now, in CHT systems, have to be continuously monitored and require the implementation of many actions to mitigate the risk of their occurrence. It shows how important it is to correctly define the context of the risk assessment to be carried out and to give new priorities to the identified risks related to the operation of operational processes by CHT systems.

5.3.4 Conclusions from the Verification of the ORA_CHTS Method

The developed framework of the risk assessment method for CHT systems is universal, following the recommendations of the ISO 31000:2018 standard. It can be applied to different operational processes (logistics and manufacturing). The process approach proposed by the Author, combining business and engineering orientations, seems to respond to contemporary challenges regarding assessing risks occurring in processes supported by CHT systems. It is because it considers the specificity of implementing operational processes in the new digitally-driven working environment, respecting the new relationship between man and technical system and the strong influence of the environment (primarily digital transformation).

However, this universal framework for the assessment procedure has been filled with guidelines, recommendations, and assumptions that consider the specifics of the internal transport process supported by CHT systems. In preparing the requirements for the various stages of the procedure, the Author drew on the results of many years of research into the impact of digital transformation on emerging risks in internal logistics systems, including transport systems in particular. This research included basic research, industrial research and development work. Therefore, the presented method is based on the literature and industrial research results, which were then verified in real logistics and transport systems. Its adaptation to industrial requirements was confirmed by the verification carried out in a selected manufacturing company, the results of which are presented in this chapter.

The experts participating in the study highly appreciated the usefulness of the method developed, the logic of the analytical procedure and the unambiguity of the assumptions made for the successive stages of the assessment. They also confirmed its comprehensiveness by assuming an interdisciplinary risk identification and analysis approach. After the experiment, critical team members recommended its use in the risk assessment process in the company under study, which is the best assessment for the verification. The experts paid particular attention to the need to prioritise identifying adverse events. This is mainly because, due to the assumed interdisciplinarity, the list of potential adverse events may be so extensive that the costs of their complete analysis will be too high and the implementation time is too long to achieve a satisfactory result. In this way, the experts confirmed the critical importance of the context identification stage for risk assessment, which was also highlighted by the Author in the prepared characterisation.

The costs associated with preparing such a comprehensive risk assessment are mainly due to the time-consuming nature of the risk identification and analysis process. However, formulated in the form of an assessment context, an unambiguous definition of the framework of the analysis and defining priorities for the identified events allow for the direct and clarifying of the scope of the analysis. This makes it possible to increase the efficiency of the assessment team. Of course, these costs and time can be reduced, including through big data and artificial intelligence. However, it is crucial to remember that these solutions will be effective primarily in Model 1 when assessing events known to managers and monitored by digital systems and responsible persons.

On the other hand, regarding unknown risks (Model 3), implementing I4.0 solutions into the risk assessment process requires much work to investigate cause-and-effect relationships and build appropriate analytical algorithms, which generates high preparation costs. In this case, the economic viability of incurring the costs of implementing such solutions concerning the potential benefits of the risk knowledge should be assessed. In many cases, using quantitative methods and expertise using Model 3 will be more effective in addressing operational risk than reaching for complex digital solutions.

An important consideration in implementing the method is correctly selecting risk assessment team members. The team should include representatives with different knowledge and experience concerning the assessed process. The interdisciplinarity of the assessment depends mainly on the diversity of competencies of the team members, who can look at the analysed process from different points of view and perceive different types of risks related to the transport operations performed. Due to the specific nature of CHT systems, the team should include a manager responsible for the results of the process and managing the team, specialists with organisational knowledge about the implementation of the process, technical knowledge related to the operation and maintenance of the technical system, and IT knowledge to increase the efficiency and security of the use of digital systems to support the operation of the process. In the author's opinion, in the case of periodic risk assessment, it is worthwhile to include an external expert in the team who will be able to detect adverse events that are "invisible" to those involved in the daily implementation of the process.

Verifying the ORA_CHTS method confirms its usefulness and effectiveness in the industry. It can be applied to material handling processes in various industrial sectors, whether in manufacturing companies, trading companies or logistics operators. Its flexible framework allows the scope of the analysis to be customised to the company's specific needs. It also does not require the use of advanced computing techniques. The choice of risk identification and analysis methods depends on the assessment team's capacity and the availability of quantitative data. It is left to the discretion of those responsible for risk management in the organisation. A critical aspect of implementing the ORA_CHTS method is introducing a risk management culture within the organisation, focused on fully understanding the benefits and risks of applying CHT systems to operational processes and the potential for using digital transformation to improve the risk assessment process.

5.4 Summary

Complex cyber-human-technical systems handle cargo transport processes in the internal logistics of smart factories. These systems are based on three pillars: (a) people, (b) transport equipment and (c) cyber systems, whose mutual cooperation determines the effectiveness and efficiency of the execution of operational processes. Today, this cooperation is taking on a different character than the systems operating the internal transport processes before the fourth industrial revolution (human-technical systems). The critical change here is the relationship between man and transport equipment. In the traditional HT system, the machine was subordinate to the human. Man made the decisions, and the machine was merely a tool in his hands. Today, machines communicate with each other and carry out human-programmed tasks, but without human participation. Thanks to artificial intelligence, they can even make decisions on their own. The quality of these decisions depends on the original algorithms developed by humans but subsequently developed by artificial intelligence. Therefore, emerging errors in the behavioural logic of these devices are more difficult to detect, and, at the same time, the scale of their impact is far greater than it was in traditional transport systems. In HT systems, the effect of an erroneous human decision affected individual devices that were only part of the technical system in operation. In Logistics 4.0 solutions, the decision errors are usually systemic in nature, meaning that they can affect all the objects belonging to the system. It is also worth noting that technological changes are simultaneously responsible for organisational changes. This makes it necessary for the methodology of the risk assessment performed for CHT systems to change regarding the approach to identifying and analysing adverse events and to develop a dedicated context for the risk assessment performed.

Chapter 4 proved a research gap regarding adapting risk assessment methods to the specifics of the currently implemented Industry 4.0 solutions in material handling processes. Therefore, the aim of Chapter 5 was to present the framework of the risk assessment method and its verification on the example of a selected material handling system. In order to understand the changes occurring in the material handling process supported by CHT systems, it was first necessary to present a comparison of the characteristics of the material handling process supported by the human-technical system versus that supported by the cyber-human-technical system. The characteristics of the two systems are presented in Sect. 5.1. This comparison made an important contribution to the formulated risk assessment context, which was the first step of the proposed method.

In Sect. 5.2, the Author proposed a new framework for risk assessment in logistics processes supported by CHT systems (ORA_CHTS). In formulating the assumptions for ORA_CHTS, the Author adopted the interdisciplinary approach formulated in Sect. 2.3 as a process approach, which combines business and engineering orientation in the analytical procedure. A critical assumption for the conducted investigation is the assumption of a focus on operational risk, which is defined in Sect. 5.2. This assumption determines the matrix describing the range of events that are the subject of the assessment. The assessment procedure has been restricted to only those events whose impact is assessed negatively. This assumption reflects the engineering orientation characteristic of risk assessment in technical systems. At the same time, the procedure methodology was based on the ISO 31000:2018 guidelines, reflecting the business orientation adopted in the assessment procedure.

The ORA_CHTS framework is based on a six-step process. Noteworthy is the high priority given to the first stage of the procedure, which concerns defining the context for the analytical procedure to be carried out. At this stage, the characteristics of the new organisation and the flow of the internal transport process are distinguished, which determines the prioritisation of the identified risks. The assessment context defined for the proposed approach implies distinguishing four layers of analysis reflecting the main sources of adverse events occurring in the processes handled by CHT systems, namely: the human factor layer, the technical layer, the digital layer and the environment. At the same time, the formulated evaluation context implies the consideration of the phases of the process management cycle in the adopted evaluation procedure. Based on the importance given to the planning phase of the process supported by the CHT system, the evaluation context implies a separate analysis of adverse events related to the planning and execution phases, which will allow for better alignment of risk management modalities. The context thus formulated is reflected in a matrix arrangement of analytical layers.

The next step is the identification of adverse events. For this stage, a procedure was designed with five analytical steps: process and decision-making analysis in the planning and execution phases of the material handling process, identification of adverse events in both phases, classification of adverse events according to the distinguished analytical layers and their categorisation according to the knowledge available about them. The procedure formulated in this way results in a set of identified adverse events grouped according to the phases of the process management cycle and the analytical layers and broken down into known and unknown risks. Such a detailed classification of the set of events allows, in the subsequent stages of the procedure, to correctly select the methods of risk analysis and to indicate the most effective ways of proceeding to reduce the level of risk.

The categorisation of adverse events into known and unknown risks allows the appropriate risk analysis method to be applied. In ORA_CHTS, the Author proposes three models of analytical procedure, which recommend selecting appropriate methods of analysis depending on the availability of data and the knowledge of the assessment team members. Due to the universal nature of the framework, the Author does not indicate the specific analysis techniques and methods that should be used for risk assessment, as this depends significantly on the specifics of the events but also the knowledge and analytical tools available to the assessment team. The assumptions presented only make recommendations on selecting groups of analytical methods that should be used in each model described. One of the proposed methods is a risk analysis based on fuzzy set theory, which can be applied to both Model 1 and Model 2. This method addresses a current limitation in the risk analysis process of CHT systems, namely the lack of comprehensive historical data on the events under investigation. This phenomenon is currently observed in many logistics processes supported by CHT systems due to the early stage of digital maturity in several enterprises. However, in the years to come, with the development of digital transformation, the fuzzy set method will be replaced by quantitative methods. This is likely due to the availability of big data and the possibility of applying machine learning or artificial intelligence techniques in the risk analysis stage.

The fourth stage of the proposed method is risk evaluation. Using only two classic risk parameters (probability and consequences) at the analysis stage allows the evaluation stage to use a popular tool for visualising risk categories in the form of a risk matrix. This tool clearly illustrates risk classification results, dividing risks into acceptable and unacceptable. This makes it possible to precisely prioritise and target actions to reduce the risk level of the adverse events analysed. The results of the risk evaluation are the basis for improvement measures, the implementation of which is the next step in the risk management process, which, however, is not described in the ORA_CHTS method. However, it is the subject of the author's current research and will be developed in the aspect of building process resilience of organisations in supply chains.

Therefore, the characterisation of the ORA_CHTS method is concluded at stage 5 on risk monitoring. This stage is critical as it runs in parallel with all the stages defined in the framework. Its purpose is to ensure that the process is carried out correctly and that the results achieved are in line with the objectives, as well as to monitor the environment in terms of its impact on the risks arising in the internal transport process. The risk monitoring system recommended by the Author for the ORA_CHTS method is based on three elements: (a) a reporting system, (b) a system of risk indicators, and (c) a system for monitoring technical equipment and cyber security systems.

The developed ORA_CHTS method was implemented for risk assessment in a selected internal transport system using an AGV system. Implementing the method in the existing system allowed to verify the correctness of the assumptions made and the challenges associated with its implementation. The assessment results, and conclusions from verifying the method are presented in Sect. 5.3. In order to obtain comprehensive conclusions, the evaluation results obtained for the CHT system were compared with the risks for the same events estimated for the traditional TH system. This provided proof of the need to change the approach to risk assessment for the internal transport process operated by CHT systems.

References

1. Smalko Z, Nowakowski T, Tubis A (2020) Framework for reliability hazard theory (in Polish). Wroclaw University of Science and Technology Publishing House, Wrocław
2. Smalko Z (2011) Introduction to cyber threat theory (in Polish). J Konbin 4:201–214
3. Tubis A (2018) Operational risk management method in road transport (in Polish). Wroclaw University of Science and Technology Publishing House, Wrocław
4. Macías CC, Reyna MR, Ruiz-Cortés A (2011) mixing Rasci matrices and BPMN together for responsibility management. In: VII Jornadas en Ciencia e Ingeniería de Servicios (JCIS'11)
5. Kacprzyk J (1986) Fuzzy sets in systems analysis (in Polish). PWN, Warszawa
6. Łachwa A (2001) The fuzzy world of sets, numbers, relations, facts, rules and decisions (in Polish). Publish House EXIT, Warszawa
7. Zadeh LA (1965) Fuzzy sets. Inf Control 8:338–353. https://doi.org/10.1016/S0019-9958(65)90241-X
8. Bayrakdaroğlu A, Yalçın N (2013) A fuzzy multi-criteria evaluation of the operational risk factors for the state-owned and privately-owned commercial banks in Turkey. Hum Ecol Risk Assess Int J 19:443–461. https://doi.org/10.1080/10807039.2013.755098
9. Tripathy DP, Ala CK (2018) Risk assessment in underground coalmines using fuzzy logic in the presence of uncertainty. J Inst Eng (India): Ser D 99:157–163. https://doi.org/10.1007/s40033-018-0154-7
10. Tubis AA, Ryczyński J, Żurek A (2021) Risk assessment for the use of drones in warehouse operations in the first phase of introducing the service to the market. Sensors 21:6713. https://doi.org/10.3390/s21206713
11. ISO 31010:2019 Risk management—risk assessment techniques
12. Łuczak J (2009) Risk assessment methods—ISO/IEC 27001 information security managament system's key element. Sci J Marit Univ Szczecin 19:63–70
13. Tourki Y, Keisler J, Linkov I (2013) Scenario analysis: a review of methods and applications for engineering and environmental systems. Environ Syst Decis. 33:3–20. https://doi.org/10.1007/s10669-013-9437-6
14. Girling P (2022) Operational risk management: a complete guide for banking and Fintech. John Wiley & Sons Inc., Hoboken, New Jersey
15. Mughal AA (2018) The art of cybersecurity: defense in depth strategy for robust protection. Int J Intell Autom Comput 1:1–20
16. Abdelghani T (2019) Implementation of defense in depth strategy to secure industrial control system in critical infrastructures. Am J Artif Intell 3:17. https://doi.org/10.11648/j.ajai.20190302.11
17. Paul B, Rao M (2022) Zero-trust model for smart manufacturing industry. Appl Sci 13:221. https://doi.org/10.3390/app13010221

Chapter 6
Conclusions and Future Research

Risk assessment is becoming increasingly important in the management of modern companies as a system to support corporate decision-making processes. Numerous scientific publications confirm its growing importance and application in developing risk assessment methods and how they can be used to improve processes in organisations and build the resilience of companies and supply chains to emerging disruptions. Department managers and process owners also express the demand for risk assessment results in improving their decisions. Risk management is no longer the domain of top management and strategic management. Nowadays, lower-level managers and process owners see the potential in its application at the operational level. The importance attached to risk assessment is also evident in the quality certification documents for the management of organisations. The ISO 9001:2015 series places much more emphasis on risk management than was included in the 9001:2008 standard. ISO auditors emphasise that it is expected that the next version of ISO 9001 will take place even more emphasis on assessing and dealing with risks identified for business operations, observing the growing importance of risk in business management. At the same time, standards and norms are being developed exclusively for risk management, such as COSO (Enterprise Risk Management—Integrated Framework and Enterprise Risk Management—Integrating Strategy with Performance); FERMA Risk Management Standard—AIRMIC/ALARM/IRM; Australian Standard AZ/ZNS 4360; PN-ISO 31000:2018 Risk Management—Principles and Guidelines; ISO 28000—Security Management System in the Supply Chain. Some of these are discussed in Chap. 2. Also, trends and phenomena emerging in the environment, such as project management, requirements to build resilient supply chains, but also the COVID-19 pandemic, increase managers' demand for risk assessment results and their applicability in tactical and operational planning processes. These trends and their relevance were also identified and described by the Author in Chap. 1. However, the phenomenon that has had the most significant impact on the writing of this book is the accelerating digital transformation, which is not only changing the technical infrastructure used to support individual operations but is strongly modifying the organisation of

A. Tubis, *The New Paradigm of Risk in Internal Transport Supporting Logistics 4.0 System*, Lecture Notes in Intelligent Transportation and Infrastructure,
https://doi.org/10.1007/978-3-031-64920-2_6

processes and impacting the organisational culture of the company. Digital transformation brings with it enormous potential for process improvement. It expands analytical capabilities to support multidimensional risk assessment in companies. However, it also generates new types of risks and changes in the risk level of events that were not previously considered critical. For this reason, the Author considered it reasonable to present the readers with a new paradigm of risks associated with logistics processes supported by CHT systems, using the internal transport process as an example.

The basic assumption for the preparation of this book was the need to adapt risk assessment methods to the requirements of modern logistics and transport processes, which are increasingly supported by CHT systems. Implementing Industry 4.0 solutions affects not only technological and organisational changes but also the risks arising in the planning and execution phase of these processes. This requires a change in the approach to risk assessment, which should be interdisciplinary and focus on the entire process flow rather than its individual components. Therefore, the monograph aims to present a new risk assessment paradigm for Logistics 4.0 systems, particularly the internal transport system. This paradigm is concerned with the operational level of risk management. It combines two approaches to risk assessment for anthropotechnical systems sup-porting process execution in modern companies—the business approach and the engineering approach. In the book, the Author proposes a new risk assessment method for Logistics 4.0 systems based on the author's concept of an interdisciplinary process approach to risk assessment, digital transformation and automation of logistics processes, including internal transport in material handling.

The developed risk assessment method was based on the following:

- theoretical research, based on which the Author identified the currently existing research gap concerning risk assessment for internal transport processes supported by CHT systems, and
- industrial and development research, which was conducted as part of research projects and industrial internships in companies implementing Industry 4.0 solutions in logistics processes.

The comprehensive literature research conducted, primarily discussed in Chaps. 2 and 4, is a critical contribution to structuring the knowledge on risk management and the risk aspects of implementing automated material handling solutions. These results summarise the theoretical research of the last seven years conducted by the Author and her team on different approaches to risk assessment and, above all, the impact of implementing Industry 4.0 solutions on the risks arising in transport processes and the methods for their assessment. In this way, the developed ORA_CHTS method fills the identified research gap and responds to the needs currently reported by industry and academia.

Key theoretical aspects presented in this book include:

- The comprehensive literature review of different approaches to enterprise risk management and, on this basis, the development of the author's approach to risk management based on an interdisciplinary process analysis.

- The comprehensive literature review of Industry 4.0 solutions, particularly Logistics 4.0 solutions, implemented to support the material handling process. This review identified a research gap, which initiated the need to develop the ORA_CHTS method.
- The development of an operational risk assessment method for logistics processes supported by cyber-human-technical systems, which is based on:
  - The new approach to formulating the risk assessment context—based on the matrix arrangement of analytical layers, including human factor, technical, digital, and environmental layers, and a division between the planning and execution phases of the process.
  - The new approach to risk identification—based on the 5-step analytical procedure and the categorisation of risks into known and unknown. This approach distinguished 16 classes of adverse events as the basis for the identification carried out.
  - The new approach to risk analysis—based on the three defined analytical models. The choice of analytical models depends on the adverse event class and historical data access.
- Development of a new approach to risk monitoring based on the system, including:
  - Customised reporting system for risk assessment.
  - Risk Indicators System.
  - System for monitoring the performance of technical equipment and the security of cyber systems.

The ORA_CHTS method was also assessed for its feasibility of implementation in a real system. Verification of the correctness of the assumptions made and the usefulness of the assessment procedure for the business environment was presented in the example of the internal transport process in a selected manufacturing company. The method was verified positively, thus confirming the effectiveness and correctness of the formulated principles for the assessment procedure. At the same time, the team of experts and managers participating in the risk assessment process highly evaluated the results obtained. It recommended the ORA_CHTS method for implementation in the enterprise at the level of operational management support. Thus, taking into account the practical aspects of the study, the author's main achievements include:

- Validation of the developed ORA_CHTS method in the existing internal transport process supported by a CHT system using an AGV solution.
- Verify the method for existing conditions of planning and execution of the process, taking into account the requirements of the operating environment of the AGV system, the existing constraints, and the characteristics of the CHT system supporting the process.
- Verify the effectiveness of the developed method as a solution to support managerial decision-making processes.
- Development of guidelines for reporting data to information systems supporting the risk assessment process.

The developed ORA_CHTS method may be subject to further development by researchers, particularly in selecting specific methods and techniques for risk identification and analysis. Theoretical research may also address the possibility of using the assessment results to improve the handling of material flows within a company and improve cooperation in the supply chain. The developed assessment method could also find wide application in manufacturing, trading, and service companies (logistics operators, distribution centres). The ORA_CHTS method can support processes concerning:

- Formulating guidelines for the risk assessment context to be developed, with particular attention to the specifics of the internal transport processes supported by CHT systems.
- Identification of risks in the internal transport process supported by CHT systems, considering 16 classes of potential adverse events.
- Risk analysis, taking into account the occurring classes of adverse events and the availability of historical data on these events, based on the proposed analytical models.
- Develop logistics process monitoring systems guidelines to support operational risk management in companies.

Summarising all the research work carried out, the following final conclusions can be drawn:

- Based on a comprehensive review of the literature on the application of Industry 4.0 solutions in logistics processes, especially in internal transport, and from the area of risk management, it can be concluded that there is currently a research gap regarding risk assessment methods at the operational level taking into account the interdisciplinary process approach and changes in logistics processes, especially internal transport, caused by their handling by CHT systems.
- The Author also confirmed this gap based on industrial research and research internships in companies implementing and applying Industry 4.0 solutions, including AGV and UAV systems.
- The interdisciplinary approach to assessing the risks associated with using CHT systems should be based on a matrix arrangement of analytical layers considering the system's components and environment, as well as the phase of the process management cycle.
- At the risk identification and analysis stage, the nature of the adverse events to be assessed should be taken into account, particularly the level of knowledge managers have about them. The level of knowledge should be assessed at the risk identification stage, and then at the analysis stage, it should enable the adaptation of an appropriate risk estimation model.
- Current development trends and industry requirements suggest the need to build a risk management process based on the concept of continuous process improvement. Therefore, the risk assessment method implemented should also influence

the risk monitoring system built. This is because the effectiveness of the monitoring process directly determines the quality of the data used in the risk assessment process, as well as the speed of the organisation's response to changing internal factors and external phenomena affecting the level of assessed risk.

Based on the results obtained from the literature review, ongoing industrial research and work on developing methods for operational risk management in the new process environment, the Author has defined further directions for her research challenges. Thus, in line with what has already been hinted at in Chap. 5, further research work will mainly focus on the following issues:

- Use the ORA_CHTS risk assessment results to improve internal service processes and build the resilience of the CHT system to disruptions occurring in the close and external environment.
- Use the ORA_CHTS risk assessment results to build the resilience of CHT systems, handling flows in logistics networks and taking resilience requirements at the supply chain level into account.
- Adapt the ORA_CHTS method to the specificities of other operational processes, especially external transport processes.
- The development of the ORA_CHTS method considers the guidelines of Thought 5.0, i.e. human-centric and sustainable development orientation.

Through such focused research work, the developed ORA_CHTS method will be able to be further developed and adapted to new processes and operational environments. It can also be used to improve logistics processes and build resilience in CHT systems, considering the resilience requirements of supply chain participants.

GPSR Compliance

The European Union's (EU) General Product Safety Regulation (GPSR) is a set of rules that requires consumer products to be safe and our obligations to ensure this.

If you have any concerns about our products, you can contact us on ProductSafety@springernature.com

In case Publisher is established outside the EU, the EU authorized representative is:

Springer Nature Customer Service Center GmbH
Europaplatz 3
69115 Heidelberg, Germany

Batch number: 10732090

Printed by Printforce, the Netherlands